CALCULU

ROBERT GHRIST

CALCULUS BLUE GUIDE

1st edition, corrected

Published by Agenbyte Press
Jenkintown PA, USA

ISBN 978-1-944655-07-5

Robert Ghrist is the Andrea Mitchell University PIK Professor of Mathematics and Electrical & Systems Engineering at the University of Pennsylvania

CONTENTS

CONTENTS

PREFACE

THIS TEXT is intended to be a guide for teaching from the Calculus BLUE Project materials. It includes brief sketches of the lecture contents, as well as notes for classroom discussions, sample assessment problems, and answers, with hints at solutions.

This resource is for teaching a contemporary course in multivariable calculus with applications over a 14-week semester. The prerequisites are a solid single-variable calculus course (to be realized as Calculus GREEN in the fullness of time). It is assumed that students use a combination of video resources, in-class discussions, and practice problems to learn and debug the material.

I was reluctant to produce any printed materials for the Calculus BLUE Project: a fully video- and electronic-text resource package was the goal *ab initio*. Feedback from faculty convinced me that a printed guide was requisite. *Adsum*.

This guide, like the videotext, is quartered.

- ▷ VOLUME I : Vectors & Matrices
- ▷ VOLUME II : Derivatives
- ▷ VOLUME III : Integrals
- ▷ VOLUME IV : Fields

Each quarter is split into three or four weeks, and each week contains:

- ▷ Materials, referencing chapters of Calculus BLUE
- ▷ Outline of topics
- ▷ Learning objectives
- ▷ Primer of material summarizing the videotext
- ▷ Discussion questions for use in the classroom
- ▷ Sample assessment problems
- ▷ Answers and hints at solutions

Nota bene, though answers may be given, it is the process that matters.

All content including artwork is by the author.

I am grateful to my students, both at Penn and elsewhere.

This book was written in difficult times. It is dedicated to E/L.

Robert Ghrist / prof-g

VOLUME I : VECTORS & MATRICES

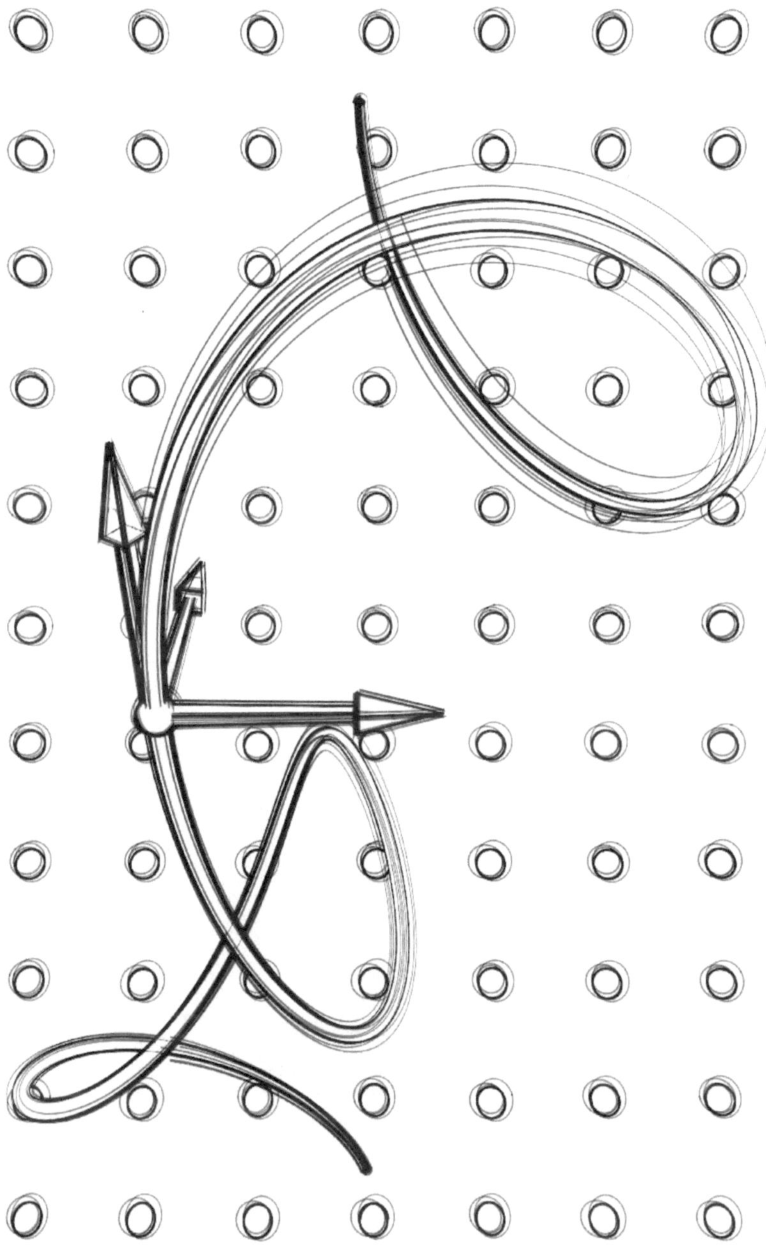

Week 1 : Points & Vectors

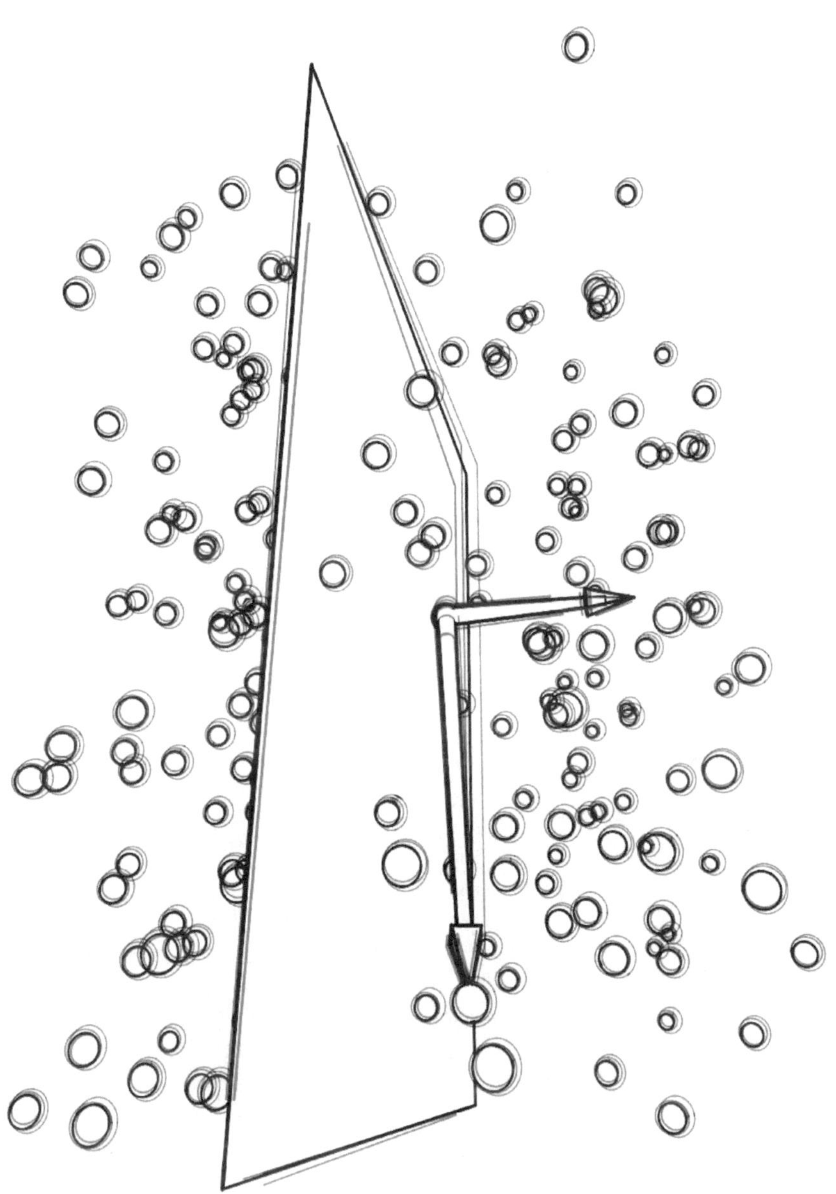

OUTLINE

MATERIALS: Calculus BLUE : Vol 1 : Chapters 1-4

TOPICS:

- ▷ Lines and planes in 2-D and 3-D
- ▷ Curves and surfaces in 2-D and 3-D
- ▷ Implicit vs. parametrized representations
- ▷ Euclidean n-dimensional space; coordinates
- ▷ Lines, planes, and hyperplanes in $\mathbb{R}^n$
- ▷ Vectors; their notation, algebra, geometry, applications

LEARNING OBJECTIVES:

- ▷ Write implicit and parametric formulae for lines and planes in 3-d
- ▷ Interpret parametric formulae for curves and surfaces in $\mathbb{R}^n$
- ▷ Use coordinates in $\mathbb{R}^n$ to compute distances between points
- ▷ Express vectors in $\mathbb{R}^n$ using coordinates or standard basis vectors
- ▷ Perform vector addition and scalar multiplication
- ▷ Compute and compare lengths of vectors

PRIMER

This first (short) week of Multivariable Calculus has no calculus whatsoever: that comes later. We will spend the first *month* in preparation for the calculus to come, with only a few small sidequests using derivatives and integrals.

Our story begins with simple elements – lines, planes, curves, and surfaces – in 2-D and 3-D, with more dimensions to follow. Certain formulae for lines (in 2-D) and planes (in 3-D) should be known / remembered from earlier courses:

$$\text{lines}: \ ax + by = c \quad or \quad \frac{x}{a} + \frac{y}{b} = 1\,,$$
$$\text{planes}: \ ax + by + cz = d \quad or \quad \frac{x}{a} + \frac{y}{b} + \frac{z}{c} = 1\,.$$

The constants (a, b, c, d) are related to how the line/plane is inclined – *cf.* slope. Such formulae are called *implicit* representations. Implicit functions can also be used to define curves in 2-D and surfaces in 3-D. The formula for a circle or a sphere of radius R at a point (x_0, y_0) in 2-D or (x_0, y_0, z_0) in 3-D should be familiar examples:

$$(x - x_0)^2 + (y - y_0)^2 = R^2 \quad : \quad (x - x_0)^2 + (y - y_0)^2 + (z - z_0)^2 = R^2\,.$$

There are numerous related formulae for quadratic surfaces in 3-D: *ellipsoids, paraboloids, hyperboloids, cones,* and more are standard examples. The formulae for such can be ignored for the time being. Eventually, you will want familiarity with spheres, ellipsoids, and cones. Many of the more unusual quadratic surfaces will not appear in our story.

Implicit representations of curves and surfaces have a mirror image in the form of *parametric* representations, where one or two parameters are used to trace out the object like so: a 2-D curve $f(t) = \big(x(t), y(t)\big)$ or a 3-D surface $g(s,t) = \big(x(s,t), y(s,t), z(s,t)\big)$. These parametrized curves and surfaces are perhaps not as familiar; they will be very useful to us, as they give explicit instructions for how to draw the object. Implicit representations are indeed implicit.

COORDINATES. We will work in the *Euclidean space* $\mathbb{R}^n$ consisting of ordered n-tuples of real numbers: $\mathbb{R}^n = \{(x_1, x_2, \dots, x_n) : x_i \in \mathbb{R}\}$. The numbers x_i used to describe a point in $\mathbb{R}^n$ are called the *coordinates* of the point. In 2-D or 3-D, we often use the more familiar (x, y) and (x, y, z) coordinates.

The distance between two points in $\mathbb{R}^n$ is via the usual Pythagorean formula:

$$d(p, q) = \sqrt{(p_1 - q_1)^2 + (p_2 - q_2)^2 + \cdots + (p_n - q_n)^2}\,.$$

Distances can be deceiving when working in higher dimensions. Any two points in a unit-radius n-dimensional ball are within a distance of two from each other. However, for a unit cube in $\mathbb{R}^n$, the antipodal corners are separated by a distance of $\sqrt{n}$: a unit cube does not fit inside a ball of diameter two (or two hundred) for sufficiently high dimensions.

Lines, planes, curves, and surfaces have higher-dimensional generalizations. Curves and surfaces are easiest to represent parametrically. Higher-dimensional objects are more easily defined implicitly: consider a single implicit equation of the form

$$\sum_{i=1}^{n} c_i x_i = 1\,,$$

where the $c_1, \dots, c_n$ are constants. This is an *affine* equation (linear plus a constant) and the geometric object it encodes is called a *hyperplane.* A hyperplane is flat like a plane in 3-D and similarly divides $\mathbb{R}^n$ into two *sides.* These (and their nonlinear generalizations) are important objects in machine learning, determining *classifiers* which separate two regions of data in $\mathbb{R}^n$.

VECTORS. Vectors can be defined, for the purposes of this course, as differences between two points in $\mathbb{R}^n$. As these differences give a sequence of n real numbers, there is temptation to conflate vectors and points: this is unavoidable. We will therefore write out the components of vectors vertically, like so:

$$\boldsymbol{v} = \begin{pmatrix} v_1 \\ v_2 \\ \vdots \\ v_n \end{pmatrix}.$$

There are many possible notations for vectors: $\boldsymbol{v}, \vec{v}, \bar{v}$, and $\underline{v}$ are common. We will use bold letters to denote vectors (though you should be prepared to see others, especially on a blackboard). Other approaches to vectors may have been shown to you in other classes: they are often defined as *arrows* or *quantities with magnitude and direction*. When you take Linear Algebra, you will learn just how precise and general the true definition of a vector is. In this course, we work with the more limited *Euclidean* (and finite dimensional!) version in which vectors are finite ordered n-tuples of *scalars* (real numbers).

Vectors, like scalars, have their own arithmetic and geometry. Vectors can be *rescaled* (to double a vector, double each component). Negative rescalings *reverse the direction* as it were. Vectors can be *added*, arithmetically by adding components, or geometrically by manipulating arrows tip-to-tail. The *zero vector* has all entries zero and is an additive identity: adding it does nothing.

The ability to add and rescale vectors leads to the notion of a *linear combination* of vectors – sums of rescaled vectors. Any vector can be decomposed as a linear combination of certain fundamental *basis vectors.* In $\mathbb{R}^3$, the standard basis vectors are unit axis-aligned vectors denoted $\hat{\imath}, \hat{\jmath}$, and $\hat{k}$ respectively, for the x-, y-, and z-axes. In $\mathbb{R}^n$, one usually denotes the standard basis vectors as $\hat{e}_i$ for $i = 1 \dots n$. These vectors have all-but-one components 0, with 1 for the ith component.

Why do we choose to do multivariable calculus in terms of vectors? They provide a very convenient data structure for geometric information. For example, in the implicit formula for a plane in $\mathbb{R}^3$, we can encode the inclination as a vector:

$$ax + by + cz = d \quad \Rightarrow \quad \boldsymbol{v} = \begin{pmatrix} a \\ b \\ c \end{pmatrix}.$$

This vector $\boldsymbol{v}$ is *orthogonal* to the plane – it meets the plane at a right angle. That is convenient, but only the beginning of the utility of vectors.

DISCUSSION

QUESTION 1. What happens of you take the equation of a line, such as $2x - 3y = 7$ and interpret it in 3-D? Why is that not a line? Why is there such a difference between implicit representations of lines in 2-D versus 3-D?

Students should understand that the equation is true for all values of z and thus gives a plane in 3-D. Subsequent question: does that single 2-variable equation determine "things" in all higher dimensions? Hyperplanes are defined by one affine equation in $\mathbb{R}^n$.

QUESTION 2. If two affine equations in 3-D give an implicit representation of a line, how can a parametric representation be generated? Specifically, given

$$3x + y - z = 4 \quad : \quad x - 2y + z = 1\,,$$

how does one parametrize this line?

If students are stuck, suggest choosing some variable to act as parameter, say $y = t$. The idea of combining equations will follow. E.g., adding the equations yields $4x - y = 5$, which allows for solving for $x(t) = (y + 5)/4$, etc.

This is a foreshadowing of row operations for solving linear systems of equations.

QUESTION 3. Where does the line $x(t) = 2t - 1$; $y(t) = 3t + 2$; $z(t) = 4t$ intersect the plane given by $4x + 3y - z = 3$? What happens if it is not a plane but a more general surface? How hard can this get?

Students should see quickly to substitute and solve. Subsequent questions help with understanding the tradeoffs between implicit and parametric representations.

QUESTION 4. Given two lines in $\mathbb{R}^2$, what is their intersection? What *could* it be? What *typically* happens? What about two lines in $\mathbb{R}^3$? A line and a plane? Two planes? What about two planes in $\mathbb{R}^4$? Can you see any patterns?

This is a challenge for many students; reassure them that it's difficult but worthwhile to reason in dimension higher than three.

QUESTION 5. In $\mathbb{R}^4$, coordinatized by (x_1, x_2, x_3, x_4), what is the intersection between the (x_1, x_2) plane and the (x_3, x_4) plane?

Students can feel both confident and totally lost here. This is a good chance to go over coordinates and their use. Try pulling back to $\mathbb{R}^3$ and asking for a definition of the (x, y) plane in terms of a set of points satisfying conditions: $\{(x, y, z) : z = 0\}$. Then, ask students what it means to intersect two sets defined by conditions.

This is a foreshadowing of intersection as logical AND, to be seen again in multivariate probability.

QUESTION 6. Compute a parametrization of the line in $\mathbb{R}^4$ which passes through the points $(1,3,-2,0)$ and $(4,2,1,-5)$.

This is a good way to pivot to vectors. After doing this, ask "How difficult was this? Was it much harder to work in 4-D than in 3-D?"

QUESTION 7. Here is a parametrization of a sphere of radius R at the origin:

$$G(s,t) = \begin{pmatrix} R\cos s \sin t \\ R \sin s \sin t \\ R\cos t \end{pmatrix}$$

How can you verify that this is indeed a sphere?

Ask students to recall what the implicit equation is & the rest should follow from them. This opens the question – how was this parametrization produced? This is a foreshadowing of spherical coordinates from Week 11.

QUESTION 8. Can you think of an example where you would want to work with points in $\mathbb{R}^{1000}$?

If students get stuck, encourage them to think in terms of digital audio, images, genetic data, neuroscience data, stock portfolios, prices for inflation estimates, climate data, etc. This is a good time to ask students about what they are interested in studying for a major and how the mathematics of this course may impact that field.

QUESTION 9. For a pair of points in $\mathbb{R}^{1000}$, is the Euclidean distance the best way to describe how far apart the points are? What else could you do?

Get students to think in terms of when images are close, or when genetic sequences are close. One could use this as an opportunity to mention Hamming distance or the metropolis metric, though these are not going to be used in this course.

QUESTION 10. We will not be dealing with infinite-dimensional vectors in this class, but consider for a moment a vector $\boldsymbol{v}$ with components v_n for $n = 1, 2, 3, \ldots$. If $v_n = 1/n$, then what is the length $|\boldsymbol{v}|$ equal to?

Under what conditions on the asymptotics of v_n would finite length be guaranteed?

QUESTION 11. What does a hyperplane in $\mathbb{R}^n$ look like? Does it matter?

This is of course ill-defined, but it's worth listening to what students come up with. This is a good opportunity to talk about classifiers in Machine Learning or linear constraints in Economics, the latter coming from an equation $\sum_i C_i x_i = K$, where $C_i > 0$ is resource cost, $x_i \geq 0$ is quantity, and $K > 0$ is total budget.

QUESTION 12. What is the volume of a unit cube in $\mathbb{R}^n$? What is the volume of the inscribed ball of radius 1? What about radius 10? What happens as $n \to \infty$?

Depending on the background of the student, they may be surprised to hear that the ball of radius 10 has volume tending (swiftly!) to zero as dimension increases. For the skeptical, ask if any portion of the cube "pokes out" from the ball as dimension increases... How many corners does the cube have? Is that where all the volume "really lives"? This is a deep set of questions that foreshadows bonus material in Week 11.

ASSESSMENT PROBLEMS

PROBLEM 1. Consider the following planes in $\mathbb{R}^3$, where C is a constant:

$$2Cx - 3y + (C+4)z = 5 \quad : \quad (C+1)x + Cy - z = 1$$

A) Assuming $C = 1$ and $z = 0$, find the point of intersection of these two planes.

B) Assuming that $C = 0$, find a vector that points along the line of intersection between these two planes.

PROBLEM 2. Consider the vectors

$$\boldsymbol{a} = \begin{pmatrix} 5 \\ 3 \\ -1 \end{pmatrix} \quad : \quad \boldsymbol{b} = \begin{pmatrix} -2 \\ 1 \\ 0 \end{pmatrix}$$

A) Write down an implicit equation for a plane passing through the point (1,2,3) and orthogonal to $\boldsymbol{a}$.

B) Write down a parametrized equation of a line passing through the point (1,2,3) and tangent to the vector $\boldsymbol{b}$.

PROBLEM 3. Consider the vectors

$$\boldsymbol{u} = \begin{pmatrix} 2 \\ -1 \\ 3 \end{pmatrix} \quad : \quad \boldsymbol{v} = \begin{pmatrix} -1 \\ 4 \\ 2 \end{pmatrix}$$

A) Write down an implicit equation for a plane passing through the point $(0,-1,2)$ and orthogonal to $\boldsymbol{u}$.

B) Write down a parametrized equation of a line passing through the point $(0,-1,2)$ and tangent to the vector $\boldsymbol{v}$.

PROBLEM 4. Consider the following parametrized plane given by

$$F\begin{pmatrix} u \\ v \end{pmatrix} = \begin{pmatrix} 2u - v \\ u + v - 3 \\ 3u + v + 1 \end{pmatrix}$$

A) Give two nonzero vectors tangent to the plane.

B) Find the point on this plane which intersects the z-axis.

PROBLEM 5. Consider the following vector in $\mathbb{R}^6$: $\boldsymbol{v} = 2\hat{e}_2 + \hat{e}_3 - \hat{e}_4 + \hat{e}_5 - 3\hat{e}_6$.

A) Compute the length of $\boldsymbol{v}$.

B) Give an example of a vector with the same length as $\boldsymbol{v}$ but which is not parallel to $\boldsymbol{v}$.

C) Is there a vector with all entries positive that is parallel to $\boldsymbol{v}$? If so, give an example; if not, explain why not.

PROBLEM 6. Which of the following five vectors in $\mathbb{R}^5$ is longest/shortest?

$$u = \begin{pmatrix} 1 \\ 1 \\ -5 \\ 0 \\ 3 \end{pmatrix} \quad v = \begin{pmatrix} -5 \\ 2 \\ 2 \\ 2 \\ 0 \end{pmatrix} \quad w = \begin{pmatrix} 0 \\ 3 \\ 5 \\ 0 \\ -2 \end{pmatrix} \quad x = \begin{pmatrix} 3 \\ 0 \\ -2 \\ 0 \\ -5 \end{pmatrix} \quad y = \begin{pmatrix} -5 \\ 0 \\ 3 \\ 2 \\ -1 \end{pmatrix}$$

PROBLEM 7. A) Write down an implicit equation of a plane which intersects the x-axis at 3; the y-axis at -2; and the z-axis at -5.

B) Give an example of a vector that is orthogonal to this plane.

PROBLEM 8. A) Write down a parametrization of a line that passes through the points $(1,3,-5)$ and $(2,4,0)$ using a parameter s.

B) Write down a parametrization of a line that passes through the points $(2,4,0)$, and $(3,0,-2)$ using a parameter t.

C) Write down a parametrization of a plane that passes through the points $(1,3,-5)$, $(2,4,0)$, and $(3,0,-2)$ using parameter s and t.

PROBLEM 9. Consider the parametrized arc given by

$$\gamma(t) = \begin{pmatrix} \cos 3t \\ 1 - \sin 2t \\ 3 + \sin 3t \\ 2 + \cos 2t \end{pmatrix} \quad : \quad 0 \le t \le \frac{\pi}{2}$$

A) Write down a vector that points from one endpoint of this arc to the other. How many such vectors are there?

B) What is the distance between the endpoints of this arc?

PROBLEM 10. For what values of constant C are the planes given by

$$3Cx + 16y + Cz = 5 \quad \& \quad 12x + Cy + 4z = 17$$

parallel to each other?

PROBLEM 11. Consider the points $P = (2,-3,5)$ and $Q = (4,1,7)$.

A) What is the distance between P and Q?

B) Give a parametrization of a line passing through both P and Q.

PROBLEM 12. Consider the line parametrized via

$$\gamma(s) = \begin{pmatrix} 2s - 1 \\ 3s + 2 \\ 4s \end{pmatrix}$$

A) Find a point where this line intersects the plane given implicitly by

$$2x - 3y + z = 10$$

B) Does this line intersect this plane orthogonally? Explain.

PROBLEM 13. Consider the plane given by $4x + 12y - 5z = 6$.

A) Give an example of a vector tangent to the plane; and a vector that is orthogonal to the plane, noting which is which.

B) Parametrize a line that is orthogonal to this plane at the point $(1, 1, 2)$.

PROBLEM 14. Write down parametrizations of a line between the following points (using a parameter t from 0 to 1):

A) From $(0,1)$ to $(3,0)$ in the plane;

B) From $(0,1,7,2,-3)$ to $(3,0,5,-4,8)$ in $\mathbb{R}^5$.

PROBLEM 15. Consider the following parametrized lines:

$$\gamma_1(t_1) = \begin{pmatrix} 3+2t_1 \\ 5+3t_1 \\ 7+4t_1 \end{pmatrix} \quad : \quad \gamma_2(t_2) = \begin{pmatrix} -1+t_2 \\ -2+2t_2 \\ 11-4t_2 \end{pmatrix}$$

A) Find the point at which these two lines intersect.

B) Write down a parametrized plane $S(t_1, t_2)$ which contains both lines.

ANSWERS & HINTS

PROBLEM 1. A) $(1,-1,0)$; B) substitute to get $-3y + 4z = 5$ and $x - z = 1$. Parametrize to get, e.g., $z = t$, $y = -\frac{5}{3} + \frac{4}{3}t$, and $x = 1 + t$. The tangent vector is $3\hat{\imath} + 4\hat{\jmath} + 3\hat{k}$.

PROBLEM 2. A) $5x + 3y - z = 8$; B) $\gamma(t) = \begin{pmatrix} 1-2t \\ 2+t \\ 3 \end{pmatrix}$

PROBLEM 3. A) $2x - y + 3z = 7$; B) $\gamma(t) = \begin{pmatrix} -t \\ 4t-1 \\ 2t+2 \end{pmatrix}$

PROBLEM 4. A) $\begin{pmatrix} 2 \\ 1 \\ 3 \end{pmatrix}$ and $\begin{pmatrix} -1 \\ 1 \\ 1 \end{pmatrix}$; B) set $2u - v = 0 = u + v - 3$; solve for $u = 1, v = 2$.

PROBLEM 5. A) 4 ; B) anything with all zeros and one 4 will do ; C) lol, no.

PROBLEM 6. lengths are $\sqrt{36}, \sqrt{37}, \sqrt{38}, \sqrt{38}, \sqrt{39}$, so $\boldsymbol{u}$ shortest, $\boldsymbol{y}$ longest.

PROBLEM 7. A) $\frac{x}{3} - \frac{y}{2} - \frac{z}{5} = 1$; B) take coefficients and multiply by 30 to get $\begin{pmatrix} 10 \\ -15 \\ -6 \end{pmatrix}$.

PROBLEM 8. A) $\gamma_1(s) = \begin{pmatrix} 2+s \\ 4+s \\ 5s \end{pmatrix}$; B) $\gamma_2(s) = \begin{pmatrix} 2+t \\ 4-4t \\ -2t \end{pmatrix}$; C) $G(s,t) = \begin{pmatrix} 2+s+t \\ 4+s-4t \\ 5s-2t \end{pmatrix}$

PROBLEM 9. A) the choices are $\begin{pmatrix} -1 \\ 2 \\ 3 \\ 1 \end{pmatrix} - \begin{pmatrix} 1 \\ 1 \\ 3 \\ 3 \end{pmatrix} = \begin{pmatrix} -2 \\ 1 \\ 0 \\ -2 \end{pmatrix}$ and its opposite ; B) $\sqrt{9} = 3$.

PROBLEM 10. solve $\frac{3C}{12} = \frac{16}{C} = \frac{C}{4}$ to get $C = 8$.

PROBLEM 11. A) $\sqrt{4 + 16 + 4} = 2\sqrt{6}$; B) one choice is $\gamma(t) = \begin{pmatrix} 2+2t \\ -3+4t \\ 5+2t \end{pmatrix}$.

PROBLEM 12. A) substitute and solve to get $s = -18$ and thus $(-37, -52, -72)$; B) no.

PROBLEM 13. A) $\begin{pmatrix} 4 \\ 12 \\ -5 \end{pmatrix}$ is orthogonal; there are many choices for a tangent, e.g., $\begin{pmatrix} -3 \\ 1 \\ 0 \end{pmatrix}$, obtained by parametrizing, say, $y = t$; and $z = 0$; B) let $\gamma(s) = \begin{pmatrix} 1+4s \\ 1+12s \\ 2-5s \end{pmatrix}$.

PROBLEM 14. A) $x = 3t, y = 1 - t$; B) $z = 7 - 2t$, and keep going... :-/

PROBLEM 15. A) solve for $t_1 = -1$ and $t_2 = 2$; B) let $S(t_1, t_2) = \begin{pmatrix} 1+2t_1+t_2 \\ 2+3t_1+2t_2 \\ 3+4t_1-4t_2 \end{pmatrix}$

Week 2 : Vector Calculus

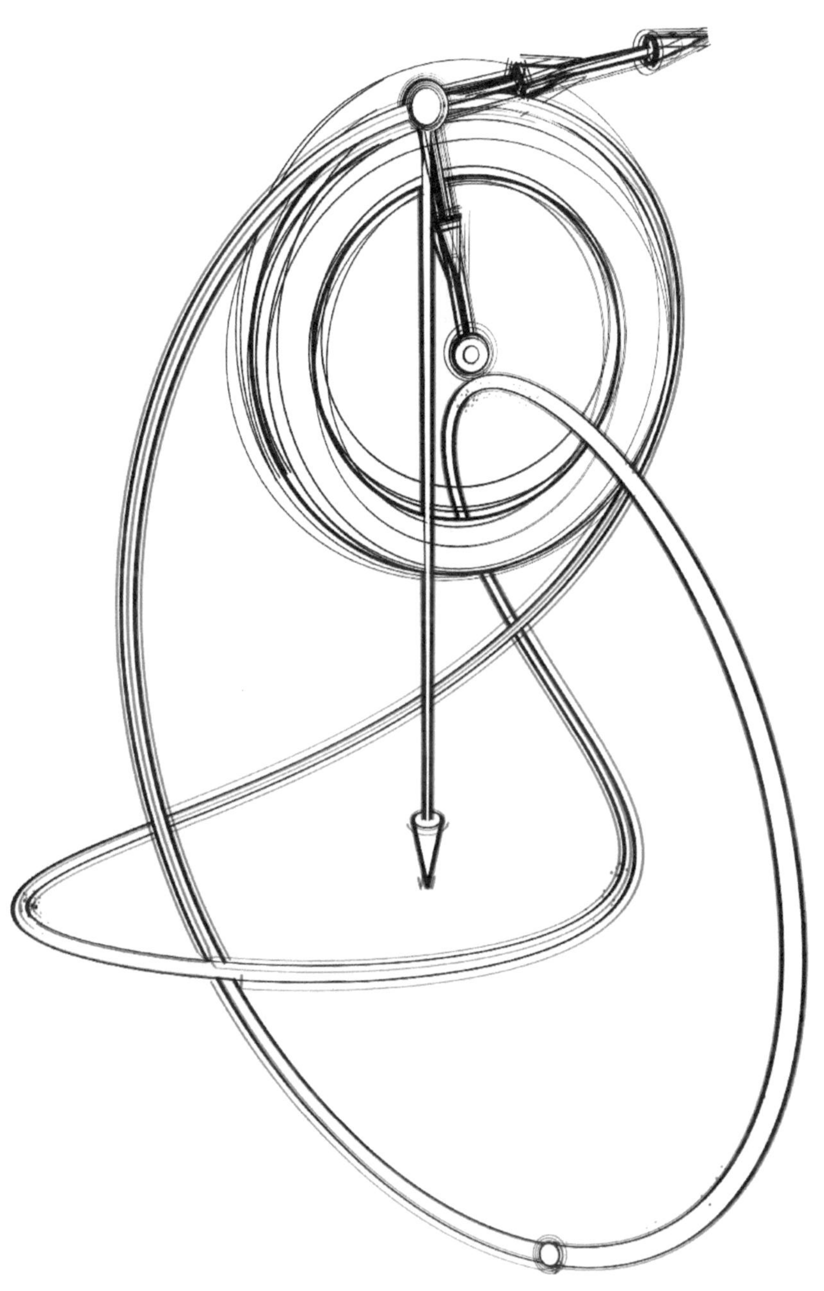

OUTLINE

MATERIALS: Calculus BLUE : Vol 1 : Chapters 5-8; Interlude

TOPICS:

- ▷ Dot product; cross product; scalar triple product
- ▷ The geometry of vectors in $\mathbb{R}^3$ and beyond
- ▷ Lengths of curves in $\mathbb{R}^n$
- ▷ Velocity and acceleration of curves
- ▷ Unit tangent and unit normal vectors to curves
- ▷ Curvature and geometry of curves
- ▷ Applications to physics of 3-D motion
- ▷ Why vector calculus of curves is not enough

LEARNING OBJECTIVES:

- ▷ Compute dot products and interpret them geometrically
- ▷ Determine angles between vectors in $\mathbb{R}^n$
- ▷ Use equations for hyperplanes in terms of dot products
- ▷ Compute cross products and interpret them geometrically
- ▷ Compute scalar triple products and interpret them geometrically
- ▷ Determine velocity and acceleration vectors of parametrized curves
- ▷ Compute unit tangent and unit normal vectors of parametrized curves
- ▷ Set up and compute arclength integrals

PRIMER

Vectors are the first step in building a language for multivariable calculus: we will use them extensively. The algebra of vectors that we saw last week extends and lifts to a version of the familiar single-variable calculus for vector-valued functions. This will not be sufficient for doing full multi-variable calculus, but it provides a good and intuitive starting point.

DOT PRODUCT. The most important operation on vectors is the *dot product*, which takes two vectors in $\mathbb{R}^n$ and returns a scalar by summing up the pairwise products of components:

$$\boldsymbol{u} \cdot \boldsymbol{v} = \sum_{i=1}^{n} u_i v_i = \boldsymbol{v} \cdot \boldsymbol{u}.$$

This has a number of nice properties, including linearity in each term. The dot product gives substantial geometric information. One begins with the observation that the dot product of a vector with itself is the square of its length: $\boldsymbol{v} \cdot \boldsymbol{v} = |\boldsymbol{v}|^2$. Any two nonzero vectors in $\mathbb{R}^n$ have a well-defined angle between them (within a plane containing both vectors). That angle, θ, is related to the dot product via:

$$\cos\theta = \frac{\boldsymbol{u} \cdot \boldsymbol{v}}{|\boldsymbol{u}||\boldsymbol{v}|}.$$

This yields a deeper and very useful interpretation. For each unit vector $\boldsymbol{u}$, the dot product $\boldsymbol{u} \cdot \boldsymbol{v}$ is the *oriented projected length* of the vector $\boldsymbol{v}$ along the $\boldsymbol{u}$-axis. Vectors are orthogonal if and only if their dot product vanishes. Positive dot products arise from vectors that point in similar directions (an acute angle).

Negative dot product connotes a more antipodal direction (an obtuse angle). We will return to this notion of an oriented projected length in Week 12.

CROSS PRODUCT. There is a fundamentally different binary operation on vectors that returns a vector – not a scalar – as the product. Unfortunately, this product is defined only in $\mathbb{R}^3$. The *cross product* of two vectors is given by the formula

$$\boldsymbol{u} \times \boldsymbol{v} = \begin{pmatrix} u_1 \\ u_2 \\ u_3 \end{pmatrix} \times \begin{pmatrix} v_1 \\ v_2 \\ v_3 \end{pmatrix} = \begin{pmatrix} u_2 v_3 - u_3 v_2 \\ u_3 v_1 - u_1 v_3 \\ u_1 v_2 - u_2 v_1 \end{pmatrix} = -(\boldsymbol{v} \times \boldsymbol{u}) .$$

This yields a vector which is orthogonal to both $\boldsymbol{u}$ and $\boldsymbol{v}$ and whose length is equal to the area of the parallelogram spanned by the two vectors, an observation that will become more and more significant over time. The geometry of the cross product is captured best by the formula for the angle between the two vectors:

$$\sin\theta = \frac{|\boldsymbol{u} \times \boldsymbol{v}|}{|\boldsymbol{u}||\boldsymbol{v}|} .$$

It is important to remember that the cross product is defined only for vectors in $\mathbb{R}^3$ (or in a plane if included in 3-D). Higher-dimensional analogues are hidden.

SCALAR TRIPLE PRODUCT. In 3-D, the dot and cross product combine into a novel operation on a triple of vectors that returns a scalar: the *scalar triple product*,

$$\boldsymbol{u} \cdot (\boldsymbol{v} \times \boldsymbol{w}) = u_1 v_2 w_3 + u_2 v_3 w_1 + u_3 v_1 w_2 - u_1 v_3 w_2 - u_2 v_1 w_3 - u_3 v_2 w_1 .$$

This has some intriguing symmetries, inherited from dot and cross products:

$$\boldsymbol{u} \cdot (\boldsymbol{v} \times \boldsymbol{w}) = \boldsymbol{v} \cdot (\boldsymbol{w} \times \boldsymbol{u}) = \boldsymbol{w} \cdot (\boldsymbol{u} \times \boldsymbol{v}) = -\,\boldsymbol{u} \cdot (\boldsymbol{w} \times \boldsymbol{v}) \ \textit{etc.}$$

The scalar triple product measures *oriented volume*: the absolute value equals the volume of the *parallelopiped* spanned by the three vectors. The sign is a type of *orientation*, to be explored in Weeks 4 and 13. For now, a nod to the right-hand rule from elementary Physics class is the best we can do to explain the +/- signs.

VECTOR CALCULUS. We have enough tools at hand to attempt building a rudimentary vector calculus. Consider a parametrized curve $\gamma: [a, b] \to \mathbb{R}^n$ as a vector-valued function. At a given $t \in [a, b]$, the value $\gamma(t)$ is a *position vector* from the origin to the point on the curve. Each coordinate x_i of γ is a single-input single-output function. As such, classical single-variable calculus tells us how to define a derivative: $\gamma' = d\gamma/dt$ is a vector-valued function whose i^{th} term equals dx_i/dt, the rate of change of the x_i coordinate with respect to t. Thinking of the parameter as time, the derivative is interpretable as a velocity vector, $\boldsymbol{v}(t)$. This velocity vector is tangent to the curve, as follows from the definition:

$$\frac{d\gamma}{dt} = \lim_{\epsilon \to 0} \frac{\gamma(t + \epsilon) - \gamma(t)}{\epsilon}$$

Note how this uses vector addition and scalar multiplication of vectors. One continues: the acceleration vector $\boldsymbol{a}(t)$ is the derivative of the velocity vector. In Physics, one tends to analyze the motion of objects along time-parametrized paths. This leads to discussions of the *unit tangent* $\hat{T}$ and *unit normal* $\hat{N}$ vectors, which are orthogonal and span the local plane of motion:

$$\hat{T} = \frac{\boldsymbol{v}}{|\boldsymbol{v}|} \quad : \quad \hat{N} = \frac{d\hat{T}/dt}{|d\hat{T}/dt|}.$$

These in turn can be used to define related quantities in Geometry, such as *curvature* and *torsion*, alluded to in the videos but not on our main storyline.

Apart from Physics, vector-valued functions are of broad interest in Mathematics, Data Science, Machine Learning, Economics, and more. One wonders: how much of multivariable calculus can be built up from properties of vectors? For example, there are some intriguing differentiation rules that mimic the product rule, but with dot products and cross products:

$$(\boldsymbol{u} \cdot \boldsymbol{v})' = \boldsymbol{u}' \cdot \boldsymbol{v} + \boldsymbol{u} \cdot \boldsymbol{v}' \quad : \quad (\boldsymbol{u} \times \boldsymbol{v})' = \boldsymbol{u}' \times \boldsymbol{v} + \boldsymbol{u} \times \boldsymbol{v}'.$$

What can be done with integration? In the same way that we differentiate a vector-valued function term-by-term, one can integrate. This is of limited use. A more helpful way to use integrals and vectors is the following computation of arclength of a parametrized curve. By the Pythagorean Theorem applied to local changes in the position, one obtains an arclength element:

$$d\ell = \left|\frac{d\gamma}{dt}\right| dt \quad : \quad \ell = \int_\gamma d\ell = \int_{t=a}^{b} \left|\frac{d\gamma}{dt}\right| dt\,.$$

This simplifies the classical formula for arclength of a graph from single variable calculus, while lifting it to arbitrary dimensions.

INTERLUDE. This arclength result – in which adding a little vector algebra to classical single-variable calculus yields a formula that works in all dimensions – is a short-lived victory. Tweaking single-variable calculus with vectors suffices for 19th and 20th century problems of projectiles and profits, satellites and sailboats. Consider the following contemporary challenges:

- ▷ Given an industrial plant manufacturing hundreds of products from thousands of components, how will output product prices change as a function of changes in input component prices?
- ▷ Coordinating a swarm of drones requires tracking positions, bearings, and velocities for every drone. If they communicate locally with neighbors, how long will it take for the swarm to align towards a goal?
- ▷ Electrical activity from hundreds of neurons are read as time series over dozens of experimental runs; how does one compute average activities and establish correlations between different neurons' activities?
- ▷ Viral infections spread based on social contact; viral propaganda spreads based on virtual contact over social networks. In what ways are epidemics and preference cascades comparable? How are their mathematical models similar or different?
- ▷ Large Language Models [LLMs] and other Neural Networks [NNs] are tuned by solving optimization problems over the relevant parameter spaces. Current LLMs have hundreds of billions of parameters.

To do contemporary multivariate calculus in full, we need more tools. This will be our focus for the next two weeks.

DISCUSSION

QUESTION 1. For what value of constant c are the planes $2cx - y + c^2z = 15$ and $x + 5cy - 3z = 4$ orthogonal? Is there more than one answer? Is there any answer?

This is a good time to remind students that when they see the word "orthogonal" they should think "dot product zero!" Does the idea of two planes being orthogonal make sense?

QUESTION 2. What is the angle between the planes $x - 2y + 3z = 6$ and $2x + 3y - z = 11$?

This again gets at the question of what the angle between planes means. In this case, it is more ambiguous: is it the "smaller" angle or the "larger" one? If confused, students should bump the problem down to two lines in a plane. Ask leading questions about the formula for computing the angle to bring students back to the meaning (and domain and codomain) of the inverse cosine function.

QUESTION 3. What is the angle between the grand diagonal of a cube in $\mathbb{R}^n$ and an incident edge?

This is a great problem to work through. Start by asking whether the orientation or size of the cube matters. Does the incident edge chosen matter? How many such edges are there? What is the angle in 2-D? in 3-D? What do you guess will happen as $n \to \infty$? What does in fact happen? If students are stuck, begin by suggesting that they assign one corner of the cube to the origin: that suggests using coordinates; the next suggestion is to recall the angle formula between two vectors.

QUESTION 4. What is the area of the triangle in the plane with vertices at (1,3), (−2,0), and (5,2)?

If stuck, help students by drawing the triangle in the plane. Is there anything that vectors can help with? It might take some imagination to get to a parallelogram cut in half.

QUESTION 5. What is the volume of the parallelopiped spanned by the vectors $\hat{\imath}$, $\hat{\jmath}$, and $\boldsymbol{v}$, where $\boldsymbol{v}$ has components v_x, v_y, and v_z?

There are several ways to do this so as to make the computations easy. After getting the (simple) final answer, think about why this makes sense geometrically. Why is it that object has this volume?

This is a foreshadowing of shearing that will be important in Week 4.

QUESTION 6. If you move along a curve in 3-D at a constant speed, what can you say about the acceleration? Is it zero?

One can of course reverse the question later and ask if the acceleration is zero, what kind of path are you travelling on?

QUESTION 7. Compute the velocity, acceleration, and arclength element of the curve with components $(\cosh t, \sinh t, \tanh t)$.

This should begin with a long digression on the hyperbolic trig functions: their definition, their properties in terms of derivatives, and perhaps their Taylor expansions. If the arclength is set up, this provides an object lesson in how very few arclength integrals can be computed in full.

QUESTION 8. Compute the length of a general helix in 3-d with radius R and height C. What are the asymptotics for small R, C? Why does this make sense?

Begin with a parametrization, making sure to place bounds on the parameter. This is a good time to recall the standard parametrization of the circle: $x(t) = R\cos 2\pi t$; $y(t) = R\sin 2\pi t$; $0 \le t \le 1$, as it will be so frequently used. What should the last coordinate be to have $0 \le z \le C$? When the answer is computed, try taking limits as different constants go to zero: as $R \to 0$, the helix becomes a straight line, but as $C \to 0$, the helix becomes a circle. Can you infer a relationship to the classic Pythagoras Theorem?

QUESTION 9. Compute the arclength of the parametrized curve in 4-D:

$$\gamma(t) = \begin{pmatrix} A\cos t \\ A\sin t \\ B\cos t \\ B\sin t \end{pmatrix} \ : \ 0 \leq t \leq 2\pi$$

This discussion should begin with trying to visualize what the curve looks like. Is it a circle? What happens when you plot this in the (x_1, x_2) plane? in the (x_2, x_4) plane? Thankfully, the arclength element is easily integrated. What are the asymptotics of the solution as $A \to 0$? This is a good problem for getting comfortable with parametrized curves in higher dimensions.

QUESTION 10. What is the projected length of the vector $\boldsymbol{w}$ onto the "$\boldsymbol{v}$-axis":

$$\boldsymbol{w} = \begin{pmatrix} 5 \\ -6 \\ 2 \\ -7 \end{pmatrix} \ : \ \boldsymbol{v} = \begin{pmatrix} 0 \\ 3 \\ 4 \\ 0 \end{pmatrix}$$

One uses the dot product, normalizing $\boldsymbol{v}$ by length. However, what does (oriented) projected length mean? Consider the dot product with $\hat{\imath}$ (or $\hat{e}_1$ in higher dimensions). We know this means "record the ith component." In fact, this very operation – take a dot product with $\hat{\imath}$, or project to the x-axis – will in Week 12 be given a new name: dx.

QUESTION 11. It was stated that the unit tangent $\hat{T}$ and the unit normal $\hat{N}$ vectors to a curve are always orthogonal, but this was never proved. Is it true? Why?

This is a good introduction to how to argue. Try to get students to start with a criterion for being orthogonal (dot product zero). How can this be shown? Let students struggle a bit with the definition of the unit normal vector as a derivative. Hint to the students that $\hat{T}$ has unit length – what does that mean in terms of the dot product? Just a little nudge to recall the product formula for derivatives and the dot product suffices to complete a proof. Close by asking the students why the dot-product rule for derivatives holds: that will come in Week 6.

QUESTION 12. What is the average length (or length-squared) of a random binary vector in n dimensions?

This is difficult/impossible to answer with the tools at hand and is meant to get students thinking and/or arguing. How can one interpret this geometrically? Is it any easier to compute the average dot product of two random binary vectors in n-D? How might these answers/guesses trend as $n \to \infty$?

ASSESSMENT PROBLEMS

PROBLEM 1. Consider the following planes in $\mathbb{R}^3$, where C is constant:

$$2Cx - 3y + (C+4)z = 5$$
$$(C+1)x + Cy - z = 1$$

A) Find the value(s) of C that makes these planes orthogonal.

B) Explain why if the planes are orthogonal at one intersection point they are orthogonal at all intersection points.

PROBLEM 2. Consider the following vectors in $\mathbb{R}^3$:

$$\boldsymbol{u} = \begin{pmatrix} 1 \\ 0 \\ -1 \end{pmatrix} \ : \ \boldsymbol{v} = \begin{pmatrix} 2 \\ -1 \\ 3 \end{pmatrix} \ : \ \boldsymbol{w} = \begin{pmatrix} -2 \\ 0 \\ 2 \end{pmatrix}$$

A) Compute the cross product $\boldsymbol{u} \times \boldsymbol{v}$.

B) Use a scalar-triple-product to compute the volume of the "parallelopiped" generated by these three vectors $\boldsymbol{u}, \boldsymbol{v}, \boldsymbol{w}$.

PROBLEM 3. Consider the following hyperplanes in $\mathbb{R}^4$:

$$\begin{aligned} x_1 - 2x_2 + 5x_3 + x_4 &= 8 \\ 3x_1 + 6x_2 + x_3 + 4x_4 &= 17 \end{aligned}$$

Identify vectors orthogonal to each hyperplane and use these to show that the hyperplanes are orthogonal.

PROBLEM 4. Consider the following planes in $\mathbb{R}^3$:

$$\begin{aligned} 2x - 3y + z &= 5 \\ 3x + y - 2z &= -1 \end{aligned}$$

Identify vectors orthogonal to each plane and use these to compute a vector that is tangent to both planes.

PROBLEM 5. Consider the following vector in $\mathbb{R}^5$:

$$\boldsymbol{v} = \begin{pmatrix} 2 \\ -1 \\ 3 \\ -1 \\ 1 \end{pmatrix}$$

A) Give an example of a nonzero vector that is orthogonal to $\boldsymbol{v}$.

B) Compute the angle between the vector $\boldsymbol{v}$ and the basis vector $\hat{e}_1$.

PROBLEM 6. Consider the following vectors in $\mathbb{R}^4$:

$$\boldsymbol{u} = \begin{pmatrix} 2 \\ 1 \\ -2 \\ 0 \end{pmatrix} \ : \ \boldsymbol{v} = \begin{pmatrix} 2 \\ 1 \\ 2 \\ -4 \end{pmatrix}$$

A) Compute the angle between these two vectors..

B) Find a vector that is orthogonal to both $\boldsymbol{u}$ and $\boldsymbol{v}$.

PROBLEM 7. Consider the following vectors in $\mathbb{R}^4$:

$$\boldsymbol{u} = \begin{pmatrix} 4 \\ -5 \\ 2 \\ -2 \end{pmatrix} \ : \ \boldsymbol{v} = \begin{pmatrix} 3 \\ 0 \\ 4 \\ 0 \end{pmatrix}$$

A) Compute the angle between these two vectors.

B) Compute the projected length of $\boldsymbol{u}$ onto the "$\boldsymbol{v}$-axis".

PROBLEM 8. Consider the following vectors in $\mathbb{R}^3$:

$$\boldsymbol{u} = \begin{pmatrix} 3 \\ -1 \\ 2 \end{pmatrix} \ : \ \boldsymbol{v} = \begin{pmatrix} -1 \\ 1 \\ 4 \end{pmatrix} \ : \ \boldsymbol{w} = \begin{pmatrix} 2 \\ 0 \\ -3 \end{pmatrix}$$

A) Without computing the angles between them, figure out which pair of vectors, above, has the largest and smallest angle between them.

B) Compute the volume of the parallelopiped spanned by $\boldsymbol{u}$ and $\boldsymbol{v}$ and $\boldsymbol{w}$. *Show work.*

PROBLEM 9. Consider the following vectors in $\mathbb{R}^4$

$$a = \begin{pmatrix} 0 \\ -1 \\ 0 \\ 2 \end{pmatrix} \; : \; b = \begin{pmatrix} 0 \\ 3 \\ -2 \\ -1 \end{pmatrix} \; : \; c = \begin{pmatrix} 1 \\ -3 \\ 2 \\ 0 \end{pmatrix} \; : \; d = \begin{pmatrix} -1 \\ 1 \\ 0 \\ 2 \end{pmatrix} \; : \; e = \begin{pmatrix} 1 \\ 2 \\ 0 \\ 4 \end{pmatrix}$$

A) Which of the above vectors have a positive dot product with $\boldsymbol{c}$?

B) Is there a pair of vectors from $\boldsymbol{a}, \boldsymbol{b}, \boldsymbol{c}, \boldsymbol{d}, \boldsymbol{e}$ that is orthogonal?

C) Give an example of a nonzero vector that is orthogonal to both $\boldsymbol{a}$ and $\boldsymbol{b}$.

PROBLEM 10. Consider the following four vectors in $\mathbb{R}^3$:

$$a = \begin{pmatrix} 1 \\ 0 \\ -3 \end{pmatrix} \; : \; b = \begin{pmatrix} 0 \\ 2 \\ 5 \end{pmatrix} \; : \; c = \begin{pmatrix} 1 \\ 3 \\ 0 \end{pmatrix} \; : \; d = \begin{pmatrix} 4 \\ 1 \\ 0 \end{pmatrix}$$

A) Is there a pair of orthogonal vectors among the above? *Explain.*

B) Which three of the vectors above span a parallelopiped with the largest volume?

PROBLEM 11. Consider the following two curves in $\mathbb{R}^3$:

$$\gamma_1(s) = \begin{pmatrix} s^2 - 3s \\ e^s - 1 \\ 1 - \cos 2s \end{pmatrix} \; : \; \gamma_2(t) = \begin{pmatrix} \sin(t-1) \\ t^2 + t - 2 \\ 1 - \sqrt{t} \end{pmatrix}$$

A) Verify that these curves intersect at the origin for some values of s and t.

B) At what angle do these curves intersect at the origin?

C) Find a vector that is orthogonal to both curves at the origin.

PROBLEM 12. Consider the parametrized curve in 3-D given by

$$\gamma(t) = \begin{pmatrix} t^2 - t + 4 \\ t^3 - 3t^2 + 2t - 1 \\ 2t \end{pmatrix} \begin{matrix} \leftarrow x \\ \leftarrow y \\ \leftarrow z \end{matrix}$$

A) Compute the velocity vector of this curve.

B) Write down the equation of a plane orthogonal to this curve at $\gamma(0)$.

C) At what angle does this curve cross the (x, y) plane $z = 0$? Explain your reasoning and give your answer as best you can without a calculator…

PROBLEM 13. Consider the parameterized surface in 3-D given by

$$S\begin{pmatrix} u \\ v \end{pmatrix} = \begin{pmatrix} 4 \\ 0 \\ 3 \end{pmatrix} + \begin{pmatrix} u^2 \\ -3u \\ 2u \end{pmatrix} + \begin{pmatrix} 0 \\ -v^2 \\ 5v \end{pmatrix}$$

Note that the point $P = (5, -4, 10)$ is on this surface (at $u = v = 1$).

A) Find two nonparallel vectors tangent to the surface at this point P.

B) Find a nonzero vector perpendicular (i.e., orthogonal) to the surface at this point P.

PROBLEM 14. Consider the parametrized curve in $\mathbb{R}^4$ given by

$$\gamma(s) = \begin{pmatrix} 3 - s \\ s^2 + 2s \\ (s+1)^{-1} \\ -s \end{pmatrix}$$

A) Compute the velocity vector to this curve at the point where $s = -2$.

B) Compute the acceleration vector to this curve at the point where $s = -2$.

C) Are the velocity and acceleration vectors at this point (where $s = -2$) orthogonal? Why or why not?

PROBLEM 15. Consider the following curve in $\mathbb{R}^3$:

$$\gamma(t) = \begin{pmatrix} (2t-1)^2 \\ t^3 - 3t + 4 \\ 3t - 2 \end{pmatrix} \quad : \quad 0 \le t \le 2$$

A) Compute the velocity vector of this curve.

B) Compute the unit tangent vector at the point where $t = 1$.

C) Set up but do not solve an integral to compute the arclength of this curve.

PROBLEM 16. Consider the following curve in $\mathbb{R}^6$:

$$\gamma(t) = \begin{pmatrix} 5\cos t \\ 4t \\ 2t - 1 \\ 5 - 2t \\ 3\sin t \\ -4\sin t \end{pmatrix} \quad : \quad 0 \le t \le \pi$$

A) Compute the velocity vector of this curve.

B) The length of the velocity vector is constant: what is it?

C) Compute the arclength of this curve in $\mathbb{R}^6$.

PROBLEM 17. Consider the following curve in $\mathbb{R}^4$, where C is constant:

$$\gamma(t) = \begin{pmatrix} 5\cos t \\ C\sin t \\ 4\cos t \\ -6\sin t \end{pmatrix} \quad : \quad 0 \le t \le \pi$$

A) Compute the velocity and acceleration vectors of this curve.

B) Compute and simply (if you can) the arclength element $d\ell$ for this curve.

C) The arclength element is constant for some value(s) of C: for what value(s)?

PROBLEM 18. Consider the parametrized curve in 3-D given by

$$\gamma(t) = \begin{pmatrix} 3 - t \\ t^3 + t - 2 \\ t^2 - 3t + 4 \end{pmatrix} \begin{matrix} \leftarrow x \\ \leftarrow y \\ \leftarrow z \end{matrix}$$

A) Compute the velocity vector of this curve and evaluate at $t = 1$.

B) Compute the acceleration vector of this curve and evaluate at $t = 1$.

C) At the point $\gamma(1) = (2,0,2)$, the velocity and acceleration vectors lie in a plane. Compute a nonzero vector that is orthogonal to this plane.

PROBLEM 19. Consider the curve given by

$$\gamma(t) = \begin{pmatrix} t^2 - 5t + 4 \\ t(t-1)(t-2) \\ e^{t-1} - 1 \end{pmatrix}$$

A) Compute the velocity and acceleration vectors to this curve and evaluate these at the point where γ intersects the origin.

B) Compute an implicit equation for a plane in $\mathbb{R}^3$ orthogonal to the curve γ at the origin in the form ______ $x +$ ______ $y +$ ______ $z =$ ______.

PROBLEM 20. Consider the parametrized path

$$\gamma(t) = \begin{pmatrix} \ln(1+t) \\ e^{3t} \\ \sin(2t) \end{pmatrix}$$

A) Compute the velocity ($\boldsymbol{v}$) and acceleration ($\boldsymbol{a}$) vectors to this curve.

B) Compute the cross product $\boldsymbol{v}(0) \times \boldsymbol{a}(0)$.

ANSWERS & HINTS

PROBLEM 1. A) set the dot product of coefficient vectors to zero to get $C^2 - C - 2 = 0$ hence $C = 2$ or $C = -1$; B) The vector orthogonal to the plane is constant, and the dot product of those two vectors determined the angle of intersection: it is constant.

PROBLEM 2. A) $\boldsymbol{u} \times \boldsymbol{v} = \begin{pmatrix} -1 \\ -5 \\ -1 \end{pmatrix}$; B) $\boldsymbol{u} \cdot (\boldsymbol{v} \times \boldsymbol{w}) = \boldsymbol{w} \cdot (\boldsymbol{u} \times \boldsymbol{v}) = \begin{pmatrix} -2 \\ 0 \\ 2 \end{pmatrix} \cdot \begin{pmatrix} -1 \\ -5 \\ -1 \end{pmatrix} = 0$

PROBLEM 3. The coefficient vectors satisfy $\begin{pmatrix} 1 \\ -2 \\ 5 \\ 1 \end{pmatrix} \cdot \begin{pmatrix} 3 \\ 6 \\ 1 \\ 4 \end{pmatrix} = 0$

PROBLEM 4. take the cross product of the orthogonal vectors to the planes to obtain $\begin{pmatrix} 2 \\ -3 \\ 1 \end{pmatrix} \times \begin{pmatrix} 3 \\ 1 \\ -2 \end{pmatrix} = \begin{pmatrix} 5 \\ 8 \\ 11 \end{pmatrix}$

PROBLEM 5. A) choose any vector with dot product zero ; B) $\arccos\left(\frac{1}{2}\right) = \frac{\pi}{3}$

PROBLEM 6. A) $\arccos(1/15)$; B) do not try to take a cross product, as it is undefined!

PROBLEM 7. A) $\arccos(4/7)$; B) $\boldsymbol{u} \cdot \boldsymbol{v}/|\boldsymbol{v}| = 4$

PROBLEM 8. A) note that $\boldsymbol{u} \cdot \boldsymbol{v} > 0$, $\boldsymbol{v} \cdot \boldsymbol{w} < 0$, and $\boldsymbol{u} \cdot \boldsymbol{w} = 0$, so that $\boldsymbol{u}, \boldsymbol{v}$ are acute and $\boldsymbol{v}, \boldsymbol{w}$ are obtuse ; B) $\boldsymbol{u} \cdot (\boldsymbol{v} \times \boldsymbol{w}) = -18$

PROBLEM 9. A) $\boldsymbol{a}$, and do not forget $\boldsymbol{c}$! ; B) no; all dot products are nonzero ; C) make up a vector with dot product zero, or take the cross product of the latter three terms in $\boldsymbol{a}, \boldsymbol{b}$ (since these lie in a 3-dimensional subspace)

PROBLEM 10. A) no ; B) scalar triple products are $\pm 9, \pm 19, \pm 33, \pm 55$, $(\boldsymbol{b}, \boldsymbol{c}, \boldsymbol{d})$ is largest

PROBLEM 11. A) let $s = 0, t = 1$; B) $\gamma_1'(0) = \begin{pmatrix} -3 \\ 1 \\ 0 \end{pmatrix}$ and $\gamma_2'(1) = \begin{pmatrix} 1 \\ 3 \\ -\frac{1}{2} \end{pmatrix}$ have dot product zero and thus are orthogonal ; C) via cross product or observation, $C\begin{pmatrix} 1 \\ 3 \\ 20 \end{pmatrix}$

PROBLEM 12. A) $\gamma'(t) = \begin{pmatrix} 2t-1 \\ t^2 - 6t + 2 \\ 2 \end{pmatrix}$; B) $2y - x + 2z = -6$; C) take the dot product of $\gamma'(0)$ with $\hat{k}$; the formula yields an angle of $\arccos\left(\frac{2}{3}\right)$ with the z-axis or, better still, an angle of $\frac{\pi}{2} - \arccos\left(\frac{2}{3}\right)$ with the (x, y) plane

PROBLEM 13. A) compute the velocity vectors with respect to u and v to obtain tangent vectors $\begin{pmatrix} 2u \\ -3 \\ 2 \end{pmatrix}$ and $\begin{pmatrix} 0 \\ -2v \\ 5 \end{pmatrix}$, with evaluation at $u = v = 1$ yieding $\begin{pmatrix} 4 \\ -3 \\ 2 \end{pmatrix}$ and $\begin{pmatrix} 0 \\ -4 \\ 5 \end{pmatrix}$; B) the cross product of these two vector yields $\begin{pmatrix} -7 \\ -20 \\ -16 \end{pmatrix}$

PROBLEM 14. A) $\gamma'(-2) = \begin{pmatrix} -1 \\ -2 \\ -1 \\ -1 \end{pmatrix}$; B) $\gamma''(-2) = \begin{pmatrix} 0 \\ 2 \\ -2 \\ 0 \end{pmatrix}$; C) $\gamma'(-2) \cdot \gamma''(-2) = -2 \neq 0$, so the vectors are not orthogonal here

PROBLEM 15. A) $\gamma'(t) = \begin{pmatrix} 4(2t-1) \\ 3(t^2-1) \\ 3 \end{pmatrix}$; B) $\hat{T}(1) = \frac{1}{5}\begin{pmatrix} 4 \\ 0 \\ 3 \end{pmatrix}$; C) the arclength is

$$\ell = \int_{t=0}^{2} \sqrt{16(2t-1)^2 + 9(t^2-1)^2 + 9}\, dt$$

PROBLEM 16. A) $\gamma'(t) = \begin{pmatrix} -5\sin t \\ 4 \\ 2 \\ -2 \\ 3\cos t \\ -4\cos t \end{pmatrix}$; B) $|\gamma'| = \sqrt{49} = 7$; C) $\ell = \int d\ell = 7\pi$

PROBLEM 17. A) $\gamma'(t) = \begin{pmatrix} -5\sin t \\ C\cos t \\ -4\sin t \\ -6\cos t \end{pmatrix}$ and $\gamma''(t) = -\gamma(t)$;

B) $d\ell = \sqrt{41\sin^2 t + (36 + C^2)\cos^2 t}\ dt$; C) choose $C = \pm\sqrt{5}$ to get $d\ell = \sqrt{41}\, dt$

PROBLEM 18. A) $\gamma' = \begin{pmatrix} -1 \\ 3t^2+1 \\ 2t-3 \end{pmatrix}$ and $\gamma'(1) = \begin{pmatrix} -1 \\ 4 \\ -1 \end{pmatrix}$; B) $\gamma'' = \begin{pmatrix} 0 \\ 6t \\ 2 \end{pmatrix}$ so $\gamma''(1) = \begin{pmatrix} 0 \\ 6 \\ 2 \end{pmatrix}$; C) it suffices to take the cross product or by observation choose $C\begin{pmatrix} 7 \\ 1 \\ -3 \end{pmatrix}$

PROBLEM 19. A) $\gamma' = \begin{pmatrix} 2t-5 \\ 3t^2-6t+2 \\ e^{t-1} \end{pmatrix}$ and $\gamma'' = \begin{pmatrix} 2 \\ 6t-6 \\ e^{t-1} \end{pmatrix}$ at origin ($t = 1$) these evaluate to $\begin{pmatrix} -3 \\ -1 \\ 1 \end{pmatrix}$ and $\begin{pmatrix} 2 \\ 0 \\ 1 \end{pmatrix}$; B) use $\gamma'(1)$ to obtain $-3x - y + z = 0$

PROBLEM 20. A) $\gamma' = \begin{pmatrix} (1+t)^{-1} \\ 3e^{3t} \\ 2\cos 2t \end{pmatrix}$ and $\gamma'' = \begin{pmatrix} -(1+t)^{-2} \\ 9e^{3t} \\ -4\sin 2t \end{pmatrix}$; B) $\begin{pmatrix} 1 \\ 3 \\ 2 \end{pmatrix} \times \begin{pmatrix} -1 \\ 9 \\ -4 \end{pmatrix} = \begin{pmatrix} -30 \\ 2 \\ 12 \end{pmatrix}$

Week 3 :
Matrix Algebra

OUTLINE

MATERIALS: Calculus BLUE : Vol 1 : Chapters 9-13

TOPICS:

- ▷ Matrices; sizes; specials (identity, zero, diagonal, triangular)
- ▷ The transpose operation on matrices
- ▷ The use of matrices as data structures
- ▷ Matrix multiplication, including matrix-vector multiplication
- ▷ Square matrices and powers
- ▷ Block matrices and their products
- ▷ Linear systems of equations: $A\boldsymbol{x} = \boldsymbol{b}$
- ▷ Row operations and row reduction
- ▷ Row reduction and back-substitution of augmented matrices
- ▷ Inverse matrices: definition, computation, and use

LEARNING OBJECTIVES:

- ▷ Identify sizes, rows, and columns of matrices
- ▷ Compute matrix-vector and matrix-matrix products
- ▷ Recognize that matrix multiplication is associative but not commutative
- ▷ Compute and work with the transpose of a matrix
- ▷ Convert linear systems of equations into the form $A\boldsymbol{x} = \boldsymbol{b}$
- ▷ Solve $A\boldsymbol{x} = \boldsymbol{b}$ via row reduction and back-substitution
- ▷ Use the formula for the inverse of a 2-x-2 matrix
- ▷ Compute the inverse of a square matrix via row reduction
- ▷ Solve $A\boldsymbol{x} = \boldsymbol{b}$ for $\boldsymbol{x}$ given the inverse A^{-1}
- ▷ Compute products, powers, and inverses of simple block matrices

PRIMER

MATRICES. An m-by-n matrix $A = (A_{ij})$ is, initially, a finite 2-dimensional array indexed by m rows and n columns. These are ubiquitous in applications. Of the many types of matrices relevant to calculus, the most important include the *square* matrices (where $m = n$), the *diagonal* matrices (which satisfy $A_{ij} = 0$ for all $i \neq j$), and the *triangular* matrices (where, for an *upper* or *lower* triangular matrix, $A_{ij} = 0$ for all $i > j$ and $i < j$ respectively). The *zero* matrix Z consists of all zeros; the *identity* matrix I is a square diagonal matrix whose diagonal entries are all one. These are both examples of *binary* matrices, whose entries are either zero or one.

Remembering the index ordering is necessary. The entry A_{ij} lies in the ith row and the jth column. A 1-by-n matrix can be thought of as a vector, and we will often conflate the two. An m-by-1 matrix can be rightly called a *row vector*.

The *transpose* of a matrix $A = (A_{ij})$ is the matrix A^T whose rows and columns are exchanged: thus, $(A^T_{ij}) = (A_{ji})$. One imagines flipping all entries across the matrix diagonal. The transpose of a vector, $\boldsymbol{v}^T$ can be thought of as a *row vector*.

MULTIPLICATION. The simplest algebraic operations on matrices mimic those of vectors - addition and scalar multiplication - acting termwise. The truly useful and interesting operation is that of *matrix multiplication*, which generalizes the dot product of vectors. For matrices A and B, the product AB is defined if the number of columns of A agrees with the number of rows of B. The formula is:

$$(AB)_{ij} = \sum_k A_{ik}B_{kj}\,.$$

This specializes to the case of matrix-vector multiplication $A\boldsymbol{v}$, where $\boldsymbol{v}$ is a vector whose size matches that of the number of columns of A. This multiplication is *associative*, meaning that $(AB)C = A(BC)$; it is not necessarily *commutative*, since $AB \neq BA$ in general (even if they are of the same size). The zero matrix Z acts like a "0" with respect to multiplication, and the identity matrix I acts like a "1" under multiplication: $AI = IA = A$ for all square A.

Understanding well how matrix multiplication operates is critical to our larger story of working with multivariate functions. From the definition, one can think of the entry $(AB)_{ij}$ of a product matrix AB as being the dot product of the ith row of A with the jth column of B. This is best illustrated as putting the product *in the corner*. In the case of matrix-vector multiplication, there is a very useful alternate way of thinking about the product. Given a vector $\boldsymbol{v} = (v_i)$, the product $A\boldsymbol{v}$ can be written as a linear combination of columns of A, weighted by the terms in $\boldsymbol{v}$. Internalizing this perspective now will be very useful.

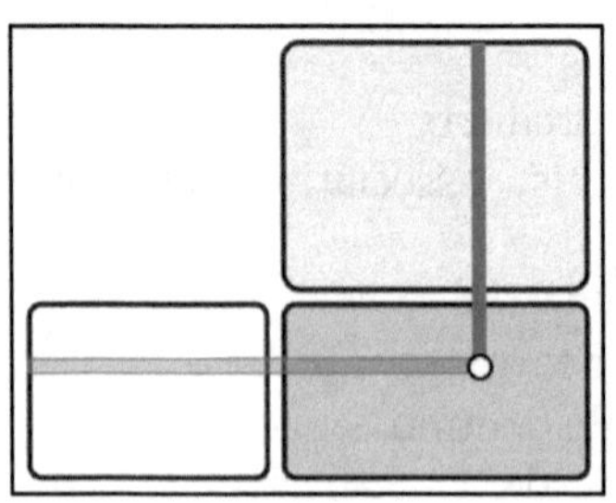

LINEAR SYSTEMS. Systems of linear (or *affine*) equations can be put into matrix-vector form as a single equation $A\boldsymbol{x} = \boldsymbol{b}$, where $\boldsymbol{x}$ is the vector of unknowns to be solved for. This type of equation is of supreme importance in applications and knowing how to solve such a system has significant implications. The high-school algebra approach of combination and substitution generalizes to a formal algorithm called *row-reduction*. There are three *row operations* which can be applied to an *augmented matrix* of the form $[\,A \mid \boldsymbol{b}\,]$ and which preserve solutions. These row operations are as follows:

- ▷ R1 : switch two rows : $R_i \leftrightarrow R_j$
- ▷ R2 : multiply a row by a nonzero scalar : $R_i \mapsto cR_i$
- ▷ R3 : combine a multiple of one row with another : $R_i \mapsto R_i + cR_j$

These operations can be performed sequentially to reduce an augmented matrix to *row-echelon form*, making the matrix as triangular as possible, with extra zeros perhaps. At any point during row reduction, one may write out the resulting system of modified equations to solve for the variables $\boldsymbol{x}$. Doing so post-reduction to row-echelon form is called *back-substitution*.

There are some subtleties to performing row-reduction. One very much wants for the upper-left corner entry (the *pivot*) to be nonzero; a value of 1 here is ideal. Note that the third row operation R3 does not rescale R_i: this will be relevant in Week 4 when computing determinants.

INVERSES. One cannot divide by a matrix to solve $A\boldsymbol{x} = \boldsymbol{b}$ as $\boldsymbol{x} = \boldsymbol{b}/A$. However, for certain square matrices A, it is possible to define something in the spirit of a reciprocal. Such an *inverse matrix* is defined to be a matrix A^{-1} such that $A^{-1}A = I = AA^{-1}$. Such can only exist in the case of a square matrix, and, even then, existence is not assured. For a 2-by-2 matrix, there is a convenient general formula:

$$A = \begin{bmatrix} a & b \\ c & d \end{bmatrix} \;:\; A^{-1} = \frac{1}{ad - bc} \begin{bmatrix} d & -b \\ -c & a \end{bmatrix}.$$

This quantity in the denominator seems to determine whether the inverse exists. When the inverse does exist, it makes short work of solving $A\boldsymbol{x} = \boldsymbol{b}$ via

$$A\boldsymbol{x} = \boldsymbol{b} \;\Rightarrow\; A^{-1}(A\boldsymbol{x}) = A^{-1}\boldsymbol{b} \;\Rightarrow\; \boldsymbol{x} = A^{-1}\boldsymbol{b}.$$

To find the inverse of a square matrix A one can apply row operations to an identity-augmented matrix $[\,A \mid I\,]$ *ad nauseum* until reducing the left side to the identity, so that the right side reveals the inverse: $[\,A \mid I\,] \to [\,I \mid A^{-1}\,]$. This is computationally tedious, but it does work, if the inverse exists. Inverses of products (assuming existence) follows a pattern reminiscent of transposes: $(AB)^{-1} = B^{-1}A^{-1}$. The few simple cases where inverse matrices can be computed without too much trouble include *triangular* matrices (where, in addition, all the diagonal terms are nonzero) and *block-diagonal* matrices with small invertible blocks along the generalized diagonal.

DISCUSSION

QUESTION 1. What are the sizes of the following matrices? Which pairwise products are well-defined?

$$A = \begin{bmatrix} 5 & 1 \\ 1 & 3 \\ -2 & 0 \end{bmatrix} \;:\; B = \begin{bmatrix} -3 & 0 & -5 \\ -1 & 7 & 1 \\ 2 & 1 & 4 \end{bmatrix} \;:\; C = \begin{bmatrix} 0 & 4 & 7 \\ 1 & -5 & 3 \end{bmatrix}$$

This question can of course be done with random matrices made up on-the-spot. Follow up with: what are the transposes of these matrices?

QUESTION 2. Can you do the following matrix-vector product in your head?

$$\begin{bmatrix} 5 & -7 & 1 & -9 \\ 1 & 3 & 0 & 11 \\ 2 & -2 & -3 & 7 \\ 0 & 5 & 1 & 2 \end{bmatrix} \begin{pmatrix} 1 \\ 0 \\ -2 \\ 0 \end{pmatrix} \quad : \quad solution = \begin{pmatrix} 5 \\ 1 \\ 2 \\ 0 \end{pmatrix} - 2 \begin{pmatrix} 1 \\ 0 \\ -3 \\ 1 \end{pmatrix}$$

Emphasize that this way of thinking in terms of linear combinations of columns will be very important next week and in the future.

QUESTION 3. Consider a pair of vectors $\boldsymbol{u}$ and $\boldsymbol{v}$ in $\mathbb{R}^n$. Can you interpret the product $\boldsymbol{u}^T\boldsymbol{v}$? How does it compare to $\boldsymbol{v}^T\boldsymbol{u}$? What about $\boldsymbol{u}\boldsymbol{v}^T$?

This helps solidify the dot product from last week, while giving a useful new notation to be seen later.

QUESTION 4. Solve the following linear system:

$$\begin{aligned} x - y + 2z &= 0 \\ 2x + y - 3z &= 1 \\ -3x + 2y + z &= 2 \end{aligned}$$

This is surely just as easy to solve via high-school algebra, but the point is not to get the solution quickly; it is to learn an algorithmic approach that will work even when the system is huge.

QUESTION 5. Let c be a constant. For what values of c does this system have a solution?

$$x + y + cz = 1$$

$$x + cy + z = 1$$
$$cx + y + z = c$$

This is a little tedious, but it communicates that solutions do not necessarily exist, and how to discern this from a row reduction.

QUESTION 6.

Recall in Week 1 Question 2 we tried to parameterize the intersection of the planes

$$3x + y - z = 4 \quad : \quad x - 2y + z = 1\,,$$

Set this up as a row-reduction problem and obtain an answer.

This is a good pattern to get into – reflecting on previous problems using new technology.

QUESTION 7. Compute the inverse of the following 3-x-3 matrix the long way, row-reducing an augmented matrix, to show

$$\begin{bmatrix} 1 & 0 & 2 \\ 3 & 1 & 0 \\ 0 & -1 & -2 \end{bmatrix}^{-1} = \frac{1}{8}\begin{bmatrix} 2 & 2 & 2 \\ -6 & 2 & -6 \\ 3 & -1 & -1 \end{bmatrix}$$

Do you think there is a general formula for computing 3-x-3 inverses? Yes. Yes, there is. Will we learn it? No, we will not.

QUESTION 8. How would you invert the following 4-by-4 matrix?

$$A = \begin{bmatrix} 2 & 5 & 0 & 0 \\ 1 & 3 & 0 & 0 \\ 0 & 0 & 5 & 3 \\ 0 & 0 & 3 & 2 \end{bmatrix} = \begin{bmatrix} B & 0 \\ 0 & C \end{bmatrix}$$

This raises the important topic of block- and block-diagonal matrices. Encourage students to figure out the properties of this structure and how it behaves under multiplication. Students may guess at the solution: how would you confirm this? Students may opt to begin with row-reduction – don't discourage that, but rather encourage them to notice what is happening structurally. Ex post, one can mention how important block-diagonal matrices are in decoupling large systems.

QUESTION 9. As it is difficult to compute an inverse of a large matrix, is there ever any justification for doing so? Is it not always better simply to row-reduce $A\mathbf{x} = \mathbf{b}$?

Try to get students to think of a situation where the b vector is changing or otherwise full of parameters that can vary. It is perhaps worth noting that explicit inverses of large matrices are almost never computed in practice: there are better algorithmic approaches.

QUESTION 10. Consider the matrix equation $AB = 0$. If these were scalars, then you know that either A or B would have to be zero. Can you find an example where this is not the case for matrices?

This will probably be too challenging. Give a hint that it can be done with a single 2-x-2 matrix whose square vanishes. A very important example of such a matrix is

$$N = \begin{bmatrix} 0 & 1 \\ 0 & 0 \end{bmatrix}.$$

This is a nilpotent matrix – its powers vanish. This matrix and matrices like it have a prominent role to play in Linear Algebra.

QUESTION 11. You can compute a square or a cube of a (square) matrix, but can you compute an arbitrary power? Try it with a diagonal matrix... Now try:

$$A = \begin{bmatrix} 2 & 1 \\ 0 & 2 \end{bmatrix}.$$

Students can perhaps guess at a pattern after a few powers. For a more principled solution (that will assist in Linear Algebra courses) remind them of the Binomial Theorem and compute A^n as

$$A^n = (2I + N)^n = (2I)^n + n(2I)^{n-1}N + \binom{n}{2}(2I)^{n-2}N^2 + \cdots$$
$$= 2^n I + n2^{n-1}N = \begin{bmatrix} 2^n & n2^{n-1} \\ 0 & 2^n \end{bmatrix}$$

This is a rather advanced problem: not for the beginner. Be sure to have done Question 10.

QUESTION 12. Why is it that the inverse of the product of two invertible matrices is the product of the inverses, but in reverse order? That is,

$$(AB)^{-1} = B^{-1}A^{-1}$$

Students should use the definition of the inverse and the associativity of multiplication.

QUESTION 13. Is it the case that the inverse of a power of an invertible matrix is the power of the inverse? That is, is it true that:

$$(A^n)^{-1} = (A^{-1})^n =: A^{-n}$$

Students should again use the definition of the inverse. This is a good opportunity to discuss the role of notation in Mathematics: we set up the notation to help us think clearly. As a follow-up question, ask what – if the notation is consistent – the definition of A^0 should be for a square matrix.

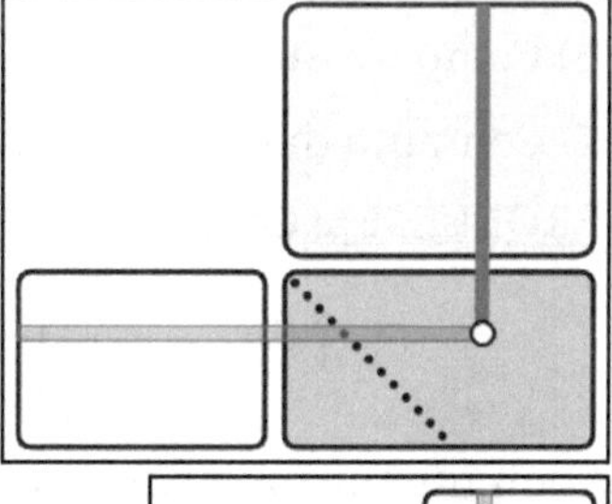

QUESTION 14. Why is it that the transpose of the product of two matrices is the product of the transposes, but in reverse order?

$$(AB)^T = B^T A^T$$

The argument used with inverses no longer works... What to do? Students will likely struggle with this one, probably resorting eventually to verifying it on an example. Point out that although this does not suffice, it does send one in the right direction. This is a good time to remind students about the power and convenience of multiplying matrices by putting the product "in the corner." Upon so doing, flipping the entire diagram along the diagonal of the product matrix reveals truth.

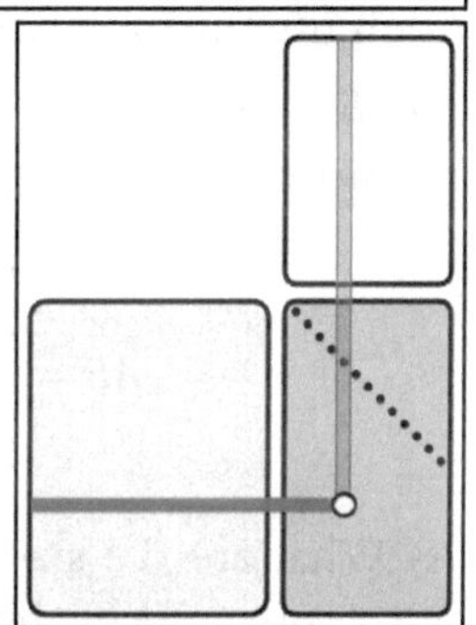

ASSESSMENT PROBLEMS

PROBLEM 1. Consider the following matrices / vectors:

$$A = \begin{bmatrix} 4 & -3 \\ 2 & 1 \end{bmatrix} \; : \; B = \begin{bmatrix} 1 & 2 & 0 \\ -1 & 4 & 3 \end{bmatrix} \; : \; \boldsymbol{u} = \begin{pmatrix} 3 \\ -2 \end{pmatrix} \; : \; \boldsymbol{v} = \begin{pmatrix} 4 \\ 1 \\ -1 \end{pmatrix}$$

Compute the following products, if possible: if not, explain why not.

A) AB B) $B\boldsymbol{v}$ C) $\boldsymbol{u}^T A$ D) $A\boldsymbol{u}$ E) A^2

PROBLEM 2. Consider the following matrices / vectors:

$$A = \begin{bmatrix} 1 & 3 \\ 2 & 2 \\ -1 & 4 \end{bmatrix} \; : \; B = \begin{bmatrix} -1 & 1 & 0 \\ 3 & -6 & 4 \\ 1 & 0 & 2 \end{bmatrix} \; : \; \boldsymbol{x} = \begin{pmatrix} 1 \\ -1 \end{pmatrix} \; : \; \boldsymbol{y} = \begin{pmatrix} 2 \\ 1 \\ 0 \end{pmatrix}$$

Compute the following products, if possible: if not, explain why not.

A) BA B) $B\boldsymbol{y}$ C) $\boldsymbol{x}^T A$ D) $A\boldsymbol{x}$ E) A^2

PROBLEM 3. Consider the following matrices / vectors:

$$A = \begin{bmatrix} 1 & -1 \\ 0 & 2 \\ -1 & 0 \end{bmatrix} : B = \begin{bmatrix} 3 & 1 & 2 \\ 5 & -1 & 2 \end{bmatrix} : \mathbf{x} = \begin{pmatrix} 2 \\ -1 \end{pmatrix} : \mathbf{y} = \begin{pmatrix} 0 \\ 1 \\ 1 \end{pmatrix}$$

Compute the following products, if possible: if not, explain why it's not possible.

A) $(BA)^2$ B) $\mathbf{y}^T A^T$ C) $(A\mathbf{x})^T$ D) $(AB)\mathbf{x}$

PROBLEM 4. Consider the following matrix and vectors:

$$A = \begin{bmatrix} 1 & -1 & 2 & 2 \\ -1 & 3 & 0 & 1 \\ 2 & 1 & -3 & 1 \\ 4 & 0 & 1 & 0 \end{bmatrix} : \mathbf{u} = \begin{pmatrix} 1 \\ 0 \\ -2 \\ 1 \end{pmatrix} : \mathbf{v} = \begin{pmatrix} 1 \\ 1 \\ 1 \\ 2 \end{pmatrix} : \mathbf{w} = \begin{pmatrix} -2 \\ -1 \\ 1 \\ -3 \end{pmatrix}$$

A) Evaluate the dot product $\mathbf{v} \cdot A\mathbf{u}$ if possible. If not, explain why not.

B) Compute $A\mathbf{u} + A\mathbf{v} + A\mathbf{w}$.

C) Compute the quantity $(\mathbf{v}\,\mathbf{w}^T)\mathbf{u}$ if it exists; if not, explain why not.

PROBLEM 5. Consider the following matrices / vectors:

$$A = \begin{bmatrix} 3 & 6 \\ -2 & 5 \\ 7 & -1 \end{bmatrix} : B = \begin{bmatrix} 4 & 1 & 3 \\ 2 & -6 & 0 \end{bmatrix} : \mathbf{x} = \begin{pmatrix} 2 \\ -1 \end{pmatrix} : \mathbf{y} = \begin{pmatrix} 3 \\ 0 \\ -2 \end{pmatrix}$$

Compute the following products, if possible: if not, explain why not.

A) AB B) $B\mathbf{y}$ C) $\mathbf{x}B$ D) $\mathbf{x}^T B\mathbf{y}$ E) A^T

PROBLEM 6. Consider the following matrix product.

$$AB = \begin{bmatrix} 1 & -4 & 0 & 7 \\ 7 & 3 & 2 & -3 \\ -4 & 1 & 4 & 8 \\ 0 & -3 & 11 & -3 \\ 3 & 2 & 0 & 1 \end{bmatrix} \begin{bmatrix} 1 & -5 & -4 & 1 & -2 \\ 2 & 5 & 4 & 3 & -3 \\ 0 & 1 & 8 & 9 & 3 \\ 3 & 2 & 2 & -1 & 8 \end{bmatrix}$$

A) What are the sizes of the two matrices A and B (write as ___-by-___) and what is the size of the product matrix AB?

B) Compute the (2,3) entry of the product matrix: $(AB)_{2,3}$.

C) Write out the transpose B^T.

PROBLEM 7. Consider the following matrix and vectors:

$$A = \begin{bmatrix} 1 & -1 & 2 & 2 \\ -1 & 3 & 0 & 1 \\ 2 & 1 & -3 & 1 \\ 4 & 0 & 1 & 0 \end{bmatrix} : \mathbf{u} = \begin{pmatrix} 2 \\ 0 \\ 2 \\ 0 \end{pmatrix} : \mathbf{v} = \begin{pmatrix} 0 \\ 1 \\ 1 \\ -1 \end{pmatrix} : \mathbf{w} = \begin{pmatrix} 1 \\ 1 \\ 1 \\ 1 \end{pmatrix}$$

A) Evaluate the dot product $\mathbf{v} \cdot A\mathbf{u}$

B) Which of the vectors, $A\mathbf{u}, A\mathbf{v}$, or $A\mathbf{w}$ is longest?

C) Compute the quantity $(\mathbf{v}\,\mathbf{w}^T)\mathbf{u}$ (if it exists ; if not, explain why not).

PROBLEM 8. Consider the following pair of linear equations:

$$3x - 5y = -6$$
$$5x - 8y = 2$$

A) Rewrite this as a linear system of the form $A\boldsymbol{x} = \boldsymbol{b}$, specifying $A, \boldsymbol{x}$, and $\boldsymbol{b}$ carefully.

B) What's the inverse of the matrix A you found above?

C) Of course, you could solve this system using basic algebra or a matrix inverse. Please do not do that. Do please solve this very carefully using row reduction and substitution.

PROBLEM 9. Consider the following row reduction of an augmented matrix:

$$\begin{bmatrix} 0 & 0 & 1 & a & 1 \\ 0 & 2 & 1 & -1 & 3 \\ 1 & -2 & 1 & 5 & c \\ 3 & b & 5 & 22 & 8 \end{bmatrix} \sim \begin{bmatrix} 1 & -2 & 1 & 5 & c \\ 0 & 2 & 1 & -1 & 3 \\ 0 & 0 & 1 & a & 1 \\ 3 & -6 & 5 & 22 & 8 \end{bmatrix} \sim \begin{bmatrix} 1 & -2 & 1 & 5 & c \\ 0 & 2 & 1 & -1 & 3 \\ 0 & 0 & 1 & a & 1 \\ 0 & 0 & 2 & 7 & 8 \end{bmatrix} \sim \begin{bmatrix} 1 & -2 & 1 & 5 & c \\ 0 & 2 & 1 & -1 & 3 \\ 0 & 0 & 1 & a & 1 \\ 0 & 0 & 0 & 3 & 6 \end{bmatrix}$$

Here, a, b, c are some constants, to be determined.

A) Using variable names x_1, x_2, x_3, x_4, write out the original system of equations as represented by the augmented matrix on the left, above. *(With a, b, c constants too...)*

B) Determine the values of the constants a, b, c with brief explanations please.

C) Using the rightmost augmented matrix above (with a, b, c values filled in), solve the system of equations.

PROBLEM 10. Consider the following row-reduction with five steps:

$$\begin{bmatrix} 3 & 13 & -5 \\ 1 & 3 & -1 \\ 2 & 4 & 3 \end{bmatrix} \xrightarrow{1} \begin{bmatrix} 1 & 3 & -1 \\ 3 & 13 & -5 \\ 2 & 4 & 3 \end{bmatrix} \xrightarrow{2} \begin{bmatrix} 1 & 3 & -1 \\ 0 & 4 & -2 \\ 2 & 4 & 3 \end{bmatrix} \xrightarrow{3}$$

$$\begin{bmatrix} 1 & 3 & -1 \\ 0 & 4 & -2 \\ 0 & -2 & 5 \end{bmatrix} \xrightarrow{4} \begin{bmatrix} 1 & 3 & -1 \\ 0 & 2 & -1 \\ 0 & -2 & 5 \end{bmatrix} \xrightarrow{5} \begin{bmatrix} 1 & 3 & -1 \\ 0 & 2 & -1 \\ 0 & 0 & 4 \end{bmatrix}$$

A) Write out descriptions of what happened at each step. Be precise, citing row numbers and what operations took place.

B) If an augmented matrix row-reduces as above to give:

$$\begin{bmatrix} 1 & 3 & -1 \\ 0 & 2 & -1 \\ 0 & 0 & 4 \end{bmatrix} \begin{pmatrix} x \\ y \\ z \end{pmatrix} = \begin{pmatrix} 13 \\ -8 \\ 24 \end{pmatrix}$$

Solve for x, y, and z.

PROBLEM 11. The following is a sequence of elementary row operations that row-reduces the beginning matrix, A, to a triangular form.

$$\begin{bmatrix} 2 & -2 & 0 & 2 & \\ 1 & -1 & 0 & -2 & 11 \\ 0 & 0 & 2 & 5 & 1 \\ 0 & -2 & 1 & 3 & -2 \\ -3 & 3 & 4 & & -15 \end{bmatrix} \Rightarrow \begin{bmatrix} 1 & -1 & 0 & 1 & 5 \\ 1 & -1 & 0 & -2 & 11 \\ 0 & 0 & 2 & 5 & 1 \\ 0 & -2 & 1 & 3 & -2 \\ -3 & 3 & 4 & & -15 \end{bmatrix} \Rightarrow \begin{bmatrix} 1 & -1 & 0 & 1 & 5 \\ 0 & 0 & 0 & -3 & 6 \\ 0 & 0 & 2 & 5 & 1 \\ 0 & -2 & 1 & 3 & -2 \\ -3 & 3 & 4 & & -15 \end{bmatrix}$$

$$\Rightarrow \begin{bmatrix} 1 & -1 & 0 & 1 & 5 \\ 0 & 0 & 0 & -3 & 6 \\ 0 & 0 & 2 & 5 & 1 \\ 0 & -2 & 1 & 3 & -2 \\ 0 & 0 & 4 & & 0 \end{bmatrix} \Rightarrow \begin{bmatrix} 1 & -1 & 0 & 1 & 5 \\ 0 & -2 & 1 & 3 & -2 \\ 0 & 0 & 2 & 5 & 1 \\ 0 & 0 & 0 & -3 & 6 \\ 0 & 0 & 4 & & 0 \end{bmatrix} \Rightarrow \begin{bmatrix} 1 & -1 & 0 & 1 & 5 \\ 0 & -2 & 1 & 3 & -2 \\ 0 & 0 & 2 & 5 & 1 \\ 0 & 0 & 0 & -3 & 6 \\ 0 & 0 & 0 & 0 & \end{bmatrix}$$

Fill in the missing entries in the matrices above.

PROBLEM 12. Consider the following system of linear equations

$$2x + 3y + 4z = 14$$

$$x + 2y + 3z = 8$$
$$4y + 7z = 9$$

A) Rewrite this as a linear system of the form $A\boldsymbol{x} = \boldsymbol{b}$, specifying $A, \boldsymbol{x}$, and $\boldsymbol{b}$ carefully.

B) Row-reduce the augmented matrix of this system to lower-triangular form.

C) Solve for the variables x, y, z using your answer to part (B).

PROBLEM 13. Consider the following row-reduction:

$$A = \begin{bmatrix} 1 & -2 & 1 & 0 \\ 3 & -4 & 8 & 3 \\ 1 & 0 & 7 & 4 \\ 0 & 2 & 5 & 6 \end{bmatrix} \Rightarrow \begin{bmatrix} 1 & -2 & 1 & 0 \\ 0 & 2 & 5 & 3 \\ 0 & 0 & 1 & 1 \\ 0 & 0 & 0 & 3 \end{bmatrix} = B$$

A) Write out the steps of the row-reduction, identifying each row operation.

B) Solve the system of equations given by

$$B\boldsymbol{x} = \begin{bmatrix} 1 & -2 & 1 & 0 \\ 0 & 2 & 5 & 3 \\ 0 & 0 & 1 & 1 \\ 0 & 0 & 0 & 3 \end{bmatrix} \begin{pmatrix} x \\ y \\ z \\ w \end{pmatrix} = \begin{pmatrix} -2 \\ 1 \\ -1 \\ 6 \end{pmatrix}$$

PROBLEM 14. Consider the following system of linear equations

$$x - 2y + 3u - v = -10$$
$$2x - 7y - 4u + v = 16$$
$$3u - 2v = -13$$
$$6u + 3v = -12$$

A) Rewrite this as a linear system of the form $A\boldsymbol{x} = \boldsymbol{b}$, specifying $A, \boldsymbol{x}$, and $\boldsymbol{b}$ carefully.

B) Row-reduce the augmented matrix of this system to lower-triangular form.

C) Solve the original equations for the unknowns using your answer to part (B) and back-substitution.

PROBLEM 15. Consider the following row-reduction of an augmented matrix:

$$\begin{bmatrix} 4 & 1 & -3 & 7 \\ 1 & 0 & -2 & 1 \\ 0 & 1 & 3 & 5 \end{bmatrix} \Rightarrow [\text{work}] \Rightarrow \begin{bmatrix} 1 & 0 & -2 & 1 \\ 0 & 1 & 5 & 3 \\ 0 & 0 & -2 & 2 \end{bmatrix}$$

A) If this comes from the augmented matrix for a system of equations $A\boldsymbol{x} = \boldsymbol{b}$, then specify the matrix A and the vector $\boldsymbol{b}$.

B) Write out the steps of the row-reduction above, identifying carefully what happens at each step.

C) Solve the original system of equations for $\boldsymbol{x} = (x \ \ y \ \ z)^T$.

PROBLEM 16. Consider the following system of linear equations

$$x_1 + 3x_2 - x_4 = -1$$
$$3x_1 + 7x_2 - 2x_3 = -5$$
$$2x_2 + 5x_3 - 3x_4 = 11$$
$$3x_3 + 5x_4 = 19$$

A) Rewrite this as a linear system of the form $A\boldsymbol{x} = \boldsymbol{b}$, specifying $A, \boldsymbol{x}$, and $\boldsymbol{b}$ carefully.

B) Row-reduce the augmented matrix of this system to lower-triangular form.

C) Solve the original equations for the unknowns using your answer to part (B).

PROBLEM 17. The matrices A and B below are invertible, with inverses given:

$$A = \begin{bmatrix} -5 & 2 & -3 \\ 4 & 2 & -3 \\ 2 & 1 & 3 \end{bmatrix} \quad : \quad A^{-1} = \frac{1}{9}\begin{bmatrix} -1 & 1 & 0 \\ 2 & 1 & 3 \\ 0 & -1 & 2 \end{bmatrix}$$

$$B = \begin{bmatrix} 2 & 3 & 4 \\ 0 & 3 & 2 \\ -2 & 5 & 1 \end{bmatrix} \quad : \quad B^{-1} = \frac{1}{2}\begin{bmatrix} 7 & -17 & 6 \\ 4 & -10 & 4 \\ -6 & 16 & -6 \end{bmatrix}$$

Solve the equation $(AB)\begin{pmatrix} x \\ y \\ z \end{pmatrix} = \begin{pmatrix} 0 \\ 1 \\ 0 \end{pmatrix}$ for x, y, and z.

PROBLEM 18. Consider the following system of linear equations

$$\begin{aligned} x_2 + 3x_3 - x_4 &= -2 \\ x_1 + 2x_2 - x_4 &= 0 \\ 3x_3 + 5x_4 &= 3 \\ x_1 + 2x_2 &= 3 \end{aligned}$$

A) Rewrite this as a linear system of the form $A\boldsymbol{x} = \boldsymbol{b}$, specifying $A, \boldsymbol{x}$, and $\boldsymbol{b}$.

B) The inverse of the matrix A you should have found above is given below. Use it to solve for the variables in the equations above.

$$A^{-1} = \begin{bmatrix} -2 & 12 & 2 & -11 \\ 1 & -6 & -1 & 6 \\ 0 & \frac{5}{3} & \frac{1}{3} & -\frac{5}{3} \\ 0 & -1 & 0 & 1 \end{bmatrix}$$

PROBLEM 19. Consider the following matrices:

$$A = \begin{bmatrix} 3 & -4 & 0 & 0 \\ 4 & -5 & 0 & 0 \\ 0 & 0 & 2 & -7 \\ 0 & 0 & -1 & 3 \end{bmatrix} \quad : \quad B = \begin{bmatrix} 0 & 0 & -2 & 4 \\ 0 & 0 & 5 & 0 \\ 2 & 1 & 0 & 0 \\ 3 & -1 & 0 & 0 \end{bmatrix}$$

A) Compute the matrix B^2

B) Compute the inverse A^{-1}

PROBLEM 20. (22c:3.2.2) In the following, please explain/follow instructions.

A) **Explain**: if A is a diagonal square matrix, then is it invertible? *If true, give reasons why; if false, give an example which demonstrates your answer.*

B) **Explain:** in the case where A and B are 3-by-3 matrices and $\boldsymbol{v}$ is a vector in $\mathbb{R}^3$, then is $A^T B\boldsymbol{v}$ also a vector in $\mathbb{R}^3$? *If true, give reasons why; if false, give an example which demonstrates your answer.*

C) **Explain**: Is it ever possible for a matrix to be its own inverse? $A^{-1} = A$ for some A? *If possible, give an example; else, give a reason why it is impossible.*

D) **Explain:** If, for a matrix B its square B^2 exists, then must B be a square matrix? *If true, give reasons why; if false, give an example which demonstrates your answer.*

PROBLEM 21. (22a:2.2.2) Consider the following square matrix

$$A = \begin{bmatrix} 1 & -4 & 0 & 0 & 0 \\ 0 & 0 & 0 & 0 & -3 \\ 0 & 0 & 4 & 0 & 0 \\ 0 & 0 & 0 & 1/3 & 0 \\ 3 & 2 & 0 & 0 & 0 \end{bmatrix}$$

Compute the inverse A^{-1} (if it exists), showing all steps.

PROBLEM 22. (22c:2.4.2) Consider the following matrices:

$$A = \begin{bmatrix} -2 & 1 \\ -1 & 0 \\ 2 & -1 \end{bmatrix} : B = \begin{bmatrix} 0 & 4 \\ 2 & 3 \end{bmatrix} : C = \begin{bmatrix} 1 & -3 \\ 0 & 2 \end{bmatrix} : D = \begin{bmatrix} 0 & 1 & 3 \\ 2 & -1 & 0 \end{bmatrix}$$

Compute/simplify the following inverses, if possible: if not, explain why not.

A) B^{-1} B) $(DA)^{-1}$ C) $(B^T)^{-1}$ D) $(CD)^{-1}$

ANSWERS & HINTS

PROBLEM 1. A) $\begin{bmatrix} 7 & -4 & -9 \\ 1 & 8 & 3 \end{bmatrix}$; B) $\begin{pmatrix} 6 \\ -3 \end{pmatrix}$; C) $[8 \quad -11]$; D) $\begin{pmatrix} 18 \\ 4 \end{pmatrix}$; E) $\begin{bmatrix} 10 & -15 \\ 10 & -5 \end{bmatrix}$

PROBLEM 2. A) $\begin{bmatrix} 1 & -1 \\ -13 & 13 \\ -1 & 11 \end{bmatrix}$; B) $\begin{pmatrix} -1 \\ 0 \\ 2 \end{pmatrix}$; C) Nope ; D) $\begin{pmatrix} -2 \\ 0 \\ -5 \end{pmatrix}$; E) Nope

PROBLEM 3. A) $\begin{bmatrix} -2 & 6 \\ -18 & 46 \end{bmatrix}$; B) Nope ; C) $[3 \quad -2 \quad -2]$; D) Nope

PROBLEM 4. A) 12 ; B) 0 ; C) $\begin{pmatrix} -7 \\ -7 \\ -7 \\ -14 \end{pmatrix}$

PROBLEM 5. A) $\begin{bmatrix} 24 & -33 & 9 \\ 2 & -32 & -6 \\ 26 & 13 & 21 \end{bmatrix}$; B) $\begin{pmatrix} 6 \\ 6 \end{pmatrix}$; C) Nope ; D) 6 ; E) $\begin{bmatrix} 3 & -2 & 7 \\ 6 & 5 & -1 \end{bmatrix}$

PROBLEM 6. A) 5-by-4 times 4-by-5 yields 5-by-5 ; B) -6 ; C) $\begin{bmatrix} 1 & 2 & 0 & 3 \\ -5 & 5 & 1 & 2 \\ -4 & 4 & 8 & 2 \\ 1 & 3 & 9 & -1 \\ -2 & -3 & 3 & 8 \end{bmatrix}$

PROBLEM 7. A) -14 ; B) $A\mathbf{u}$; C) $\begin{pmatrix} 0 \\ 4 \\ 4 \\ -4 \end{pmatrix}$

PROBLEM 8. A) $\begin{bmatrix} 3 & -5 \\ 5 & -8 \end{bmatrix}\begin{pmatrix} x \\ y \end{pmatrix} = \begin{pmatrix} -6 \\ 2 \end{pmatrix}$; B) $\begin{bmatrix} 3 & -5 \\ 5 & -8 \end{bmatrix}^{-1} = \begin{bmatrix} -8 & 5 \\ -5 & 3 \end{bmatrix}$

PROBLEM 9. B) $a = 2,\ b = -6,\ c = 0$; C) $x_1 = 1,\ x_2 = 4,\ x_3 = -3,\ x_4 = 2$

PROBLEM 10. A) steps are 1: $R_1 \Leftrightarrow R_2$, 2: $R_2 \Rightarrow R_2 - 3R_1$, 3: $R_3 \Rightarrow R_3 - 2R_1$, 4: $R_2 \Rightarrow \frac{1}{2}R_2$, 5: $R_3 \Rightarrow R_3 + R_2$; B) $z = 6,\ y = -1,\ x = 22$

PROBLEM 11. in order, (10, 7), (7), (7), (10), (10), (−2)

PROBLEM 12. A) $\begin{bmatrix} 2 & 3 & 4 \\ 1 & 2 & 3 \\ 0 & 4 & 7 \end{bmatrix}\begin{pmatrix} x \\ y \\ z \end{pmatrix} = \begin{pmatrix} 14 \\ 8 \\ 9 \end{pmatrix}$; B) $x = 3, y = 4, z = -1$

PROBLEM 13. A) $R_2 \Rightarrow R_2 - 3R_1$, then $R_3 \Rightarrow R_3 - R_1$, $R_3 \Rightarrow R_3 - R_2$, and $R_4 \Rightarrow R_4 - R_2$; B) $x = 11, y = 5, z = -3, w = 2$

PROBLEM 14. A) $\begin{bmatrix} 1 & -2 & 3 & -1 \\ 2 & -7 & -4 & 1 \\ 0 & 0 & 3 & -2 \\ 0 & 0 & 6 & 3 \end{bmatrix}\begin{pmatrix} x \\ y \\ u \\ v \end{pmatrix} = \begin{pmatrix} -10 \\ 16 \\ -13 \\ -12 \end{pmatrix}$; B) $\begin{pmatrix} x \\ y \\ u \\ v \end{pmatrix} = \begin{pmatrix} 1 \\ 0 \\ -3 \\ 2 \end{pmatrix}$

PROBLEM 15. A) $\begin{bmatrix} 4 & 1 & -3 \\ 1 & 0 & -2 \\ 0 & 1 & 3 \end{bmatrix}\begin{pmatrix} x \\ y \\ z \end{pmatrix} = \begin{pmatrix} 7 \\ 1 \\ 5 \end{pmatrix}$; C) $z = -1, y = 8, x = -1$

PROBLEM 16. A) $\begin{bmatrix} 0 & 3 & 0 & -1 \\ 3 & 7 & -2 & 0 \\ 0 & 2 & 5 & -3 \\ 0 & 0 & 3 & 5 \end{bmatrix}\begin{pmatrix} x_1 \\ x_2 \\ x_3 \\ x_4 \end{pmatrix} = \begin{pmatrix} -1 \\ -5 \\ 11 \\ 19 \end{pmatrix}$; B) $\begin{pmatrix} x_1 \\ x_2 \\ x_3 \\ x_4 \end{pmatrix} = \begin{pmatrix} -2 \\ 1 \\ 3 \\ 2 \end{pmatrix}$

PROBLEM 17. compute $B^{-1}A^{-1}\begin{pmatrix} 0 \\ 1 \\ 0 \end{pmatrix} = B^{-1}\frac{1}{9}\begin{pmatrix} 1 \\ 1 \\ -1 \end{pmatrix} = \frac{1}{9}\begin{pmatrix} -8 \\ -7 \\ 8 \end{pmatrix}$

PROBLEM 18. A) $\begin{bmatrix} 0 & 1 & 3 & -1 \\ 1 & 2 & 0 & -1 \\ 0 & 0 & 3 & 5 \\ 1 & 2 & 0 & 0 \end{bmatrix}\begin{pmatrix} x_1 \\ x_2 \\ x_3 \\ x_4 \end{pmatrix} = \begin{pmatrix} -2 \\ 0 \\ 3 \\ 3 \end{pmatrix}$; B) $\begin{pmatrix} 4 \\ 2 \\ 0 \\ 0 \end{pmatrix} + 3\begin{pmatrix} -9 \\ 5 \\ -\frac{4}{3} \\ 1 \end{pmatrix} = \begin{pmatrix} 1 \\ 17 \\ -4 \\ 3 \end{pmatrix}$

PROBLEM 19. A) $B^2 = \begin{bmatrix} 8 & -6 & 0 & 0 \\ 10 & 5 & 0 & 0 \\ 0 & 0 & 1 & 8 \\ 0 & 0 & -11 & 12 \end{bmatrix}$; B) $A^{-1} = \begin{bmatrix} -5 & 4 & 0 & 0 \\ -4 & 3 & 0 & 0 \\ 0 & 0 & -3 & -7 \\ 0 & 0 & -1 & -2 \end{bmatrix}$

PROBLEM 20. A) No ; B) Yes ; C) Yes ; D) Yes

PROBLEM 21. $A^{-1} = \begin{bmatrix} \frac{1}{7} & 0 & 0 & 0 & \frac{2}{7} \\ -\frac{3}{14} & 0 & 0 & 0 & \frac{1}{14} \\ 0 & 0 & \frac{1}{4} & 0 & 0 \\ 0 & 0 & 0 & 3 & 0 \\ 0 & -\frac{1}{3} & 0 & 0 & 0 \end{bmatrix}$

PROBLEM 22. A) $\frac{1}{8}\begin{bmatrix} -3 & 4 \\ 2 & 0 \end{bmatrix}$; B) $\begin{bmatrix} 2 & 3 \\ 3 & 5 \end{bmatrix}$; C) $\frac{1}{8}\begin{bmatrix} -3 & 2 \\ 4 & 0 \end{bmatrix}$; D) Nope

Week 4: Linear Transformations

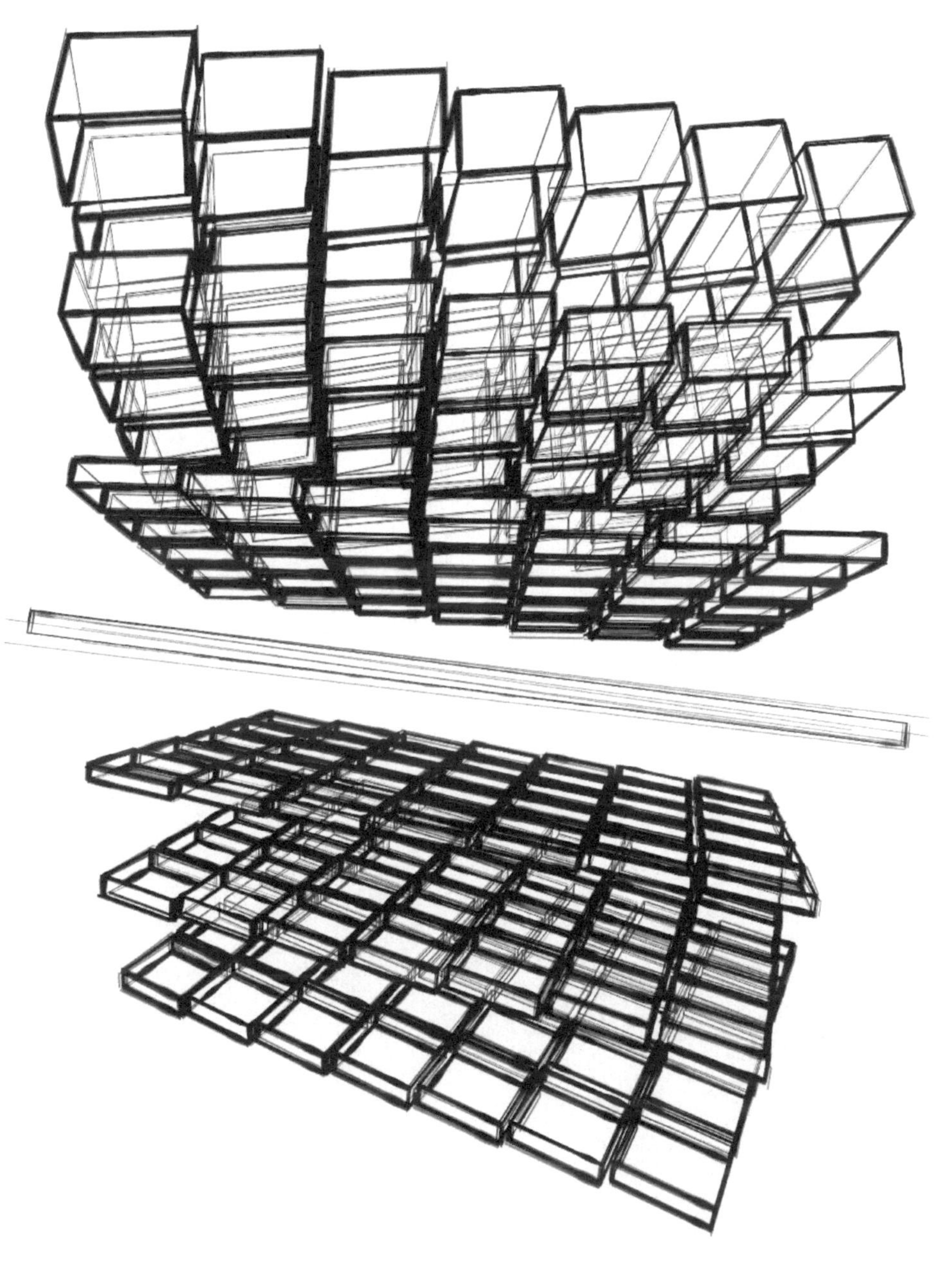

OUTLINE

MATERIALS: Calculus BLUE : Vol 1 : Chapters 14-18

TOPICS:

- ▷ Bases, including orthogonal and orthonormal bases
- ▷ Coordinates of a Euclidean vector in a given basis
- ▷ Change of coordinates from the standard basis to another
- ▷ Linear transformations; algebraic and geometric interpretations
- ▷ Rotation, rescaling, and shear matrices in 2-D
- ▷ Order of operations in linear transformations
- ▷ Determinants: computation via minor expansion
- ▷ Geometric interpretation of determinants as oriented volumes
- ▷ Computation of determinants via row reduction operations
- ▷ Determinants under products and transposes

LEARNING OBJECTIVES:

- ▷ Distinguish between general, orthogonal, and orthonormal bases
- ▷ Express a vector in coordinates of a new basis via linear system
- ▷ Recognize rotations, shears, and rescalings in terms of matrices
- ▷ Discern a linear transformation based on how it acts on basis vectors
- ▷ Compose linear transformations via matrix multiplication
- ▷ Compute determinants via minor expansion and/or blocks
- ▷ Compute volumes via determinants
- ▷ Compute determinants via row operations and reduction
- ▷ Compute determinants via matrix multiplication/factoring

PRIMER

This week marks a change from algebra to geometry; from matrices as passive data structures to matrices as active functions, transforming vectors.

BASES. We have worked with vectors in $\mathbb{R}^n$ as n-tuples of coordinates. In reality, vectors are a bit more complicated than this. Implicit in our conventions is the so-called *standard basis*: either $\hat{\imath}, \hat{\jmath}, \hat{k}$ in 3-D or $\{\hat{e}_i\}$ in $\mathbb{R}^n$. What if you had a vector of x, y, and z coordinates which you needed to send to someone whose basis was not the same as yours? Perhaps their convention was to switch the y and z axes: how would you communicate and agree upon what your vectors represent?

In the general case, consider a *basis* on $\mathbb{R}^n$ as a collection of n vectors $\boldsymbol{u}_1, \ldots, \boldsymbol{u}_n$ such that any vector $\boldsymbol{v}$ can be written *uniquely* as a linear combination of basis vectors:

$$\boldsymbol{v} = c_1\boldsymbol{u}_1 + c_2\boldsymbol{u}_2 + \cdots + c_n\boldsymbol{u}_n = \sum_{i=1}^{n} c_i\boldsymbol{u}_i ,$$

where the constants c_i are the *coefficients* of $\boldsymbol{v}$ in the new basis. There are subtleties here – how do we know this decomposition *exists* and is *unique*? This will be answered when you take linear algebra; for now, assume that you are given a basis that satisfies these conditions. Given the vector v in standard coordinates and the list of new vectors forming a basis, the coordinates c_i can be solved for by solving the linear system $A\boldsymbol{c} = \boldsymbol{v}$ for $\boldsymbol{c} = (c_i)$, where A is the matrix whose columns are the new basis vectors $\boldsymbol{u}_i$.

Certain types of bases are more or less special/convenient. A basis is said to be *orthogonal* if every pair of basis vectors is orthogonal (dot product zero). If, in addition, the basis vectors all have unit length, then the basis is called *orthonormal*.

Changing from the standard basis to a new basis is but one example of a more general phenomenon – that of transforming vectors. In the calculus to come, we will work with very general changes in coordinates: for the time being, we will focus on transformations of vectors that are *linear*.

LINEAR TRANSFORMATIONS. An m-by-n matrix A can be interpreted as a function that takes vectors as inputs and returns vectors as outputs. Consider the function $f : \mathbb{R}^n \to \mathbb{R}^m$ is defined by $f(\boldsymbol{x}) = A\boldsymbol{x}$. Any such function is a *linear transformation*, meaning that it takes the sum of two inputs to the sum of the two outputs and a rescaled input to a rescaled output:

$$f(\boldsymbol{x}+\boldsymbol{y}) = f(\boldsymbol{x}) + f(\boldsymbol{y}) \quad : \quad f(c\boldsymbol{x}) = cf(\boldsymbol{x})$$

Every linear transformation from $\mathbb{R}^n$ to $\mathbb{R}^m$ can be so represented by a m-by-n matrix: matrices are verbs as well as nouns. It is this role of a matrix as a linear function that foreshadows the primacy of matrices in multivariable calculus. Before dealing with nonlinear multivariate functions, one should understand the linear counterparts.

It is simplest to work at first with linear transformations having two inputs and two outputs: such are represented as 2-by-2 matrices. The following three examples of matrices give three fundamental types of linear transformations of the plane:

$$A = \begin{bmatrix} \lambda_1 & 0 \\ 0 & \lambda_2 \end{bmatrix} \quad : \quad B = \begin{bmatrix} \cos\theta & -\sin\theta \\ \sin\theta & \cos\theta \end{bmatrix} \quad : \quad C = \begin{bmatrix} 1 & c \\ 0 & 1 \end{bmatrix}$$

The first, A, is a *rescaling* matrix which rescales the x-axis by a factor λ_1 and the y-axis by a factor λ_2. For $|\lambda_i| > 1$, the axis is stretched; when $|\lambda_i| < 1$, the axis is squeezed. A flip happens when a coefficient λ_i is negative. The second example, B, is called a *rotation* matrix: it rotates the plane about the origin by an angle θ (in the counterclockwise direction) measured from the x-axis. The last example, C, is least familiar: it is a *shear* matrix which, in this case, preserves the x-axis and shears along the horizontal direction: the positive y-axis is sheared to the right; the negative to the left.

To understand what shears, rotations, or any other matrices do, the following is a simple but effective method. Feeding the ith basis vector $\hat{e}_i$ into A returns the ith column of A, clearly. This means that the unit cube in the domain is sent to the parallelopiped in the image spanned by the columns of A. Linearity means that the grid of cubes in the domain is sent to its image as a grid of parallelopipeds spanned by columns.

More interesting linear transformations can be obtained by composing elementary pieces. One must be careful: the linear transformation with matrix AB does not correspond to "*do A then do B*" but rather the reverse, as can be seen by associativity: $(AB)\boldsymbol{v} = A(B\boldsymbol{v})$.

DETERMINANTS. To each square matrix is associated a particular scalar that determines whether it is invertible. We encountered this *determinant* last week:

$$\det\begin{bmatrix} a & b \\ c & d \end{bmatrix} = ad - bc\,.$$

The matrix is invertible if and only if the determinant is nonzero. For a 1-by-1 matrix, the determinant is even simpler: it is the single entry. For a 3-by-3 matrix, the determinant is complicated but strangely familiar: it equals the scalar triple of the three column vectors in order. The determinant of an n-by-n matrix generalizes the scalar triple product, both algebraically and geometrically.

We noted that a linear transformation maps the standard basis vectors to the columns of the matrix. For a square matrix, the associated linear transformation maps the unit n-dimensional cube spanned by the basis vectors to a parallelopiped spanned by the column vectors. The determinant of the matrix is, precisely, the n-dimensional volume of this parallelopiped, with a plus-or-minus sign depending on an orientation (*cf.* the antisymmetry of the cross product in Week 2). This makes sense with respect to the obstruction to invertibility - smashing the unit cube to a lower-dimensional object is a violence that cannot be undone.

The geometric approach to determinants reveals a deep connection with matrices as linear transformations. Since the matrix AB represents doing B first and then A; then, for square matrices, one can interpret the actions on volumes to obtain the following fundamental result:

$$\det(AB) = (\det \mathrm{A})(\det B)\,.$$

This will be important to us much later, in Week 11 and beyond.

COMPUTATION. Computing the determinant of a general n-by-n matrix is nontrivial. There are a few ways to proceed algorithmically. One popular approach is a reductive method called *minor expansion*. Fix a matrix A; the (i,j) *minor* M_{ij} is the matrix obtained from A by deleting the ith row and the jth column. Minors of a square matrix are themselves square and have a proper determinant. The determinant of A by expansion about the ith row or the jth column (where the choice of row or column is arbitrary) is computed as an alternating sum:

$$\det A = \sum_{k=1}^{n} (-1)^{i+k} A_{ik} \det M_{ik} = \sum_{k=1}^{n} (-1)^{j+k} A_{kj} \det M_{kj}\,.$$

Minor expansion is not helpful for large matrices, since the number of operations can be factorial in n; however, for not-too-large matrices, or for matrices where one row or column has many zeros, this is a decent approach, as long as one is very careful with the alternating signs.

From this, one sees that certain matrices are trivial to work with. For a triangular matrix (upper or lower), iterated minor expansion reveals that the determinant is the product of the particular values along the diagonal - a critical result. This, in combination with the multiplicative property of the determinant and our understanding of linear transformations, leads to the capstone result of this Volume. As seen in Week 3, we can row-reduce a square matrix to a diagonal matrix via the three row reduction operations. By writing out the row operations as linear transformations (!) and computing the determinants of these three simple types of matrices, we arrive at the following conclusions:

- ▷ R1 : exchanging rows multiplies the determinant by a factor of -1;
- ▷ R2 : rescaling a row by c changes the determinant by a factor of c;
- ▷ R3 : adding one rescaled row to another leaves the determinant the same.

This is an excellent approach to determinant computation: the repeated use of the third row operation is particularly nice, as one does not have to keep track of factors.

This approach can be interpreted as factoring a matrix into a simple (triangular) term times a sequence of simple row-operation matrices. This perspective is a deep idea in applied linear algebra – matrix factorization is an extensive subject. In the context of our story, the immediate payoff is a clear proof of an otherwise difficult result to grasp: $\det A^T = \det A$.

DISCUSSION

QUESTION 1. Given the vector $3\hat{\imath} - 4\hat{\jmath}$, what is its coordinates in a new basis given by $\boldsymbol{u} = 2\hat{\imath} + 5\hat{\jmath}$ and $\boldsymbol{v} = 3\hat{\imath} + 7\hat{\jmath}$?

Emphasize the fact that the coordinates of these vectors as given are coordinates in the standard basis. Encourage students to rewrite the problem in the form of a linear system & use what they know about the inverse of a 2x2 matrix. As a follow-up, what is the general formula for the coordinates of a vector with standard coordinates c_x, c_y?

QUESTION 2. Consider the rotation matrix R_θ that rotates the plane CCW by an angle θ. How do you remember the signs on the off-diagonal terms? How do you rotate it CW instead? What are the determinant and inverse of this, and why does this make sense? What happens if you compose or take powers of rotation matrices?

There are so many good, simple questions here with rotation matrices that intersect with this week's material. This is where thinking of the columns as the image of the standard basis vectors really pays off. Be sure to spend a lot of time thinking through these with students.

QUESTION 3. What is the difference between a shear and a rotation? How many different types of shears are there? Why is it that a shear matrix does not change area? Do shears commute with rotations?

This is a good time to talk about the physical interpretation of a transpose, which flips vertical and horizontal shears. Lots of good questions that can be asked here.

QUESTION 4. What matrix corresponds to the linear transformation given by "do A then do B then do C"?

This is a good time to review associativity and commutativity. Why is it that composition of linear transformations seems to be "in the wrong order" when multiplying matrices? Always keep in mind that matrices act on vectors from the left, so think: $A\boldsymbol{x}$ then $B(A\boldsymbol{x}) = (BA)\boldsymbol{x}$ etc.

QUESTION 5. Draw two parallelograms in the plane, each with one corner at the origin and the other corner points at integer coefficients. What is the linear transformation that takes the first parallelogram to the second one?

This is challenging as stated, since it is a general 2-D change-of-basis problem in disguise. After letting students wrestle with this a bit – maybe setting it up as a linear system – try suggesting *the simpler problem of which linear transformations A and B take the unit square to each of these two shapes. After doing so, try to get students to hit on the idea of BA^{-1} as the transformation that undoes A then does B. This shows the power of composition.*

QUESTION 6. Let $\{\boldsymbol{v}_1, \ldots, \boldsymbol{v}_n\}$ denote an orthonormal basis for $\mathbb{R}^n$. If you use these vectors as columns of a square matrix Q, you have (by definition) an *orthogonal matrix* (with apologies for the confusing terminology). What is its determinant? How would you invert an orthogonal matrix Q? Is the product of two orthogonal matrices still orthogonal?

Lots of good questions to explore. The inverse problem, of course, seems too general to be done explicitly: but that is a clue. Ask students what does it mean to have an orthonormal basis. Recall from Week 3 Question 3 the relationship between the dot product and multiplication-by-transpose. This is enough to get students to see that $Q^TQ = I$. But what about QQ^T?

QUESTION 7. A *permutation matrix* is a square binary matrix such that each row and each column has exactly one "1" in it – all the rest are zeros. What can you say about the geometry of such linear transformations? What can you say about their determinants (and thus invertibility)? How would you compute an inverse?

For determinants, this permits several approaches. Try to get students to think about different ways of computing these. Are all such matrices rotations of one form or another? Are they orthogonal matrices? This might be a good time to foreshadow notions of orientability. As a follow-up question, try to get students to argue whether permutations are closed under composition.

QUESTION 8. Compute the determinant of the following matrix:

$$A = \begin{bmatrix} 0 & 0 & 0 & 1 & -2 \\ 0 & 1 & 5 & 2 & 3 \\ 0 & 0 & 7 & 0 & -6 \\ 0 & 0 & 0 & -3 & 5 \\ 2 & 4 & -7 & 1 & 3 \end{bmatrix}$$

Which way is easier? Minor expansion? Row reduction? There's room for debate here; it might be a good idea to have the class split into teams trying it different ways and then doing a compare-and-contrast. How is it that signs/orientations manifest in each approach?

QUESTION 9. For which value(s) of C are these vectors coplanar?

$$\boldsymbol{u} = \begin{pmatrix} 1 \\ 0 \\ 3C \end{pmatrix} \quad : \quad \boldsymbol{v} = \begin{pmatrix} C \\ 2 \\ -3 \end{pmatrix} \quad : \quad \boldsymbol{w} = \begin{pmatrix} 0 \\ 1 \\ C \end{pmatrix}$$

This can be viewed in terms of the scalar triple product or, perhaps better, the determinant. Why is it that having a vanishing determinant means that all three vectors lie in some plane? This can lead to good discussions.

QUESTION 10. Consider a simple block-diagonal matrix of the form

$$D = \begin{bmatrix} A & 0 \\ 0 & B \end{bmatrix}$$

Where A and B are square matrices, and the off-diagonal blocks are all zero. Argue that $\det D = (\det A)(\det B)$.

This is good practice for reasoning about determinants. If students get stuck, try suggesting thinking in terms of row reduction to triangular forms. If they solve the problem by row reduction, suggest a redo using composition & see if they can factor M as a product of two matrices where the A and B blocks are swapped out for identities. Does the order of multiplication matter?

QUESTION 11. As a follow-up to Question 10, compute the determinant of

$$C = \begin{bmatrix} 0 & B \\ A & 0 \end{bmatrix}$$

Students may be tempted to solve this via pattern-matching, without paying attention to the delicacies of signs. This is a good opportunity to teach care and precision in reasoning.

QUESTION 12. Consider the *Vandermonde* matrix V on n variables $x_1, \ldots, x_n$ whose entries are given by $V_{ij} = x_i^j$. Write out an example in the case $n = 3$ using x, y, z variables. These matrices are useful in digital signal processing, interpolation, number theory, and much more. Here is a fact about their determinants:

$$\det V = \prod_{1 \le i < j \le n} (x_j - x_i)$$

(Recall that the Π means *product*...) Can you derive this result in the $n = 3$ case? Which method(s) of computation will work best?

If students try minor expansion, go with it and see how the factorization problem presents itself. Row reduction is also not without difficulties. This is a good example of multiple pathways to compute a difficult determinant. For a more advanced student, it offers a springboard to an induction proof.

ASSESSMENT PROBLEMS

PROBLEM 1. Consider the following pair of vectors in $\mathbb{R}^2$

$$\boldsymbol{a} = 4\hat{\imath} + 6\hat{\jmath} \quad : \quad \boldsymbol{b} = 3\hat{\imath} - 2\hat{\jmath}$$

A) These form a basis for $\mathbb{R}^2$. Is this basis orthogonal? orthonormal? neither?

B) Compute the coefficients of the vector $\boldsymbol{v} = 5\hat{\imath} + 7\hat{\jmath}$ in this $(\boldsymbol{a}, \boldsymbol{b})$ basis using matrices and vectors. Identify your final answer as a- and b-coordinates.

C) What are the coefficients of the vector $\boldsymbol{a} + \boldsymbol{b}$ in this $(\boldsymbol{a}, \boldsymbol{b})$ basis?

PROBLEM 2. Consider the following three linear transformations of the plane, represented as matrices:

$$A = \begin{bmatrix} 1/2 & 0 \\ 0 & -1 \end{bmatrix} : B = \begin{bmatrix} 1 & 0 \\ -1 & 1 \end{bmatrix} : C = \begin{bmatrix} 0 & -2 \\ 2 & 0 \end{bmatrix}$$

A) Describe in words what the linear transformation B does to the plane.

B) What is the matrix that represents "first do A then do B then do C". Please express your answer as a single matrix.

C) Which vector is sent by B to the vector $\begin{pmatrix} 1 \\ 1 \end{pmatrix}$?

PROBLEM 3. Consider the linear transformation $f: \mathbb{R}^3 \to \mathbb{R}^3$ given by $f(\boldsymbol{x}) = A\boldsymbol{x}$, where

$$\boldsymbol{x} = \begin{pmatrix} x \\ y \\ z \end{pmatrix} \quad : \quad A = \begin{bmatrix} \frac{1}{2} & -\frac{\sqrt{3}}{2} & 0 \\ \frac{\sqrt{3}}{2} & \frac{1}{2} & 0 \\ 0 & 0 & -2 \end{bmatrix}$$

A) Where does this linear transformation send the point $x = 4, y = 2, z = 3$?

B) *Describe in words:* what does this linear transformation do to the z-axis?

C) *Describe in words*: what does this linear transformation do to the (x, y)-plane?

PROBLEM 4. Let A and B denote linear transformations from the plane to itself. That is, $A: \mathbb{R}^2 \to \mathbb{R}^2$ and $B: \mathbb{R}^2 \to \mathbb{R}^2$. Assume that:

1. The matrix representing A is $A = \begin{bmatrix} 2 & 1 \\ 4 & 3 \end{bmatrix}$.
2. The transformation AB takes the $\hat{\imath}$ vector to $\begin{pmatrix} 4 \\ 10 \end{pmatrix}$ and the $\hat{\jmath}$ vector to $\begin{pmatrix} -2 \\ 1 \end{pmatrix}$.

A) What is the matrix that represents the linear transformation AB?

B) What is the matrix that represents the linear transformation B?

PROBLEM 5. Consider the linear transformation $f: \mathbb{R}^2 \to \mathbb{R}^2$ given by $f(\boldsymbol{x}) = A\boldsymbol{x}$, where A is the product of three matrices:

$$\boldsymbol{x} = \begin{pmatrix} x \\ y \end{pmatrix} \quad : \quad A = \begin{bmatrix} 0 & -1 \\ 1 & 0 \end{bmatrix} \begin{bmatrix} 2 & 0 \\ 0 & -1/2 \end{bmatrix} \begin{bmatrix} 0 & 1 \\ -1 & 0 \end{bmatrix}$$

A) Describe *in words* what this linear transformation does:

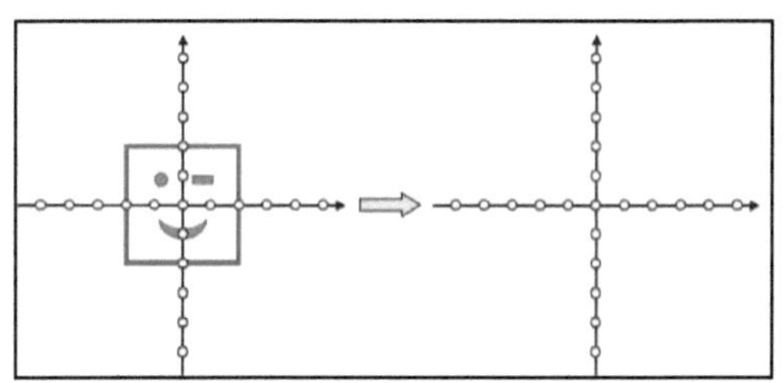

First, f does this; Next, f does this;

Finally, f does this.

B) Write out A as a single 2-by-2 matrix:

C) Draw a picture of what the linear transformation given by A does to the image on the left...

PROBLEM 6. Consider the linear transformation from $\mathbb{R}^4$ to $\mathbb{R}^4$:

$$f\begin{pmatrix} x \\ y \\ u \\ v \end{pmatrix} = \begin{pmatrix} -2y \\ 2x \\ -6u + 4v \\ 3u - 2v \end{pmatrix}$$

A) What matrix A represents this linear transformation? That is, for what matrix A is $f(\boldsymbol{x}) = A\boldsymbol{x}$, where $\boldsymbol{x} = (x \;\; y \;\; u \;\; v)^T$?

B) Do the columns of A form an orthogonal basis?

C) Describe in words what the linear transformation above does to the (x, y) plane.

PROBLEM 7. Consider the following pair of vectors in $\mathbb{R}^2$

$$\boldsymbol{u} = 5\hat{\imath} - 7\hat{\jmath} \quad : \quad \boldsymbol{v} = 4\hat{\imath} - 6\hat{\jmath}$$

A) These form a basis for $\mathbb{R}^2$. Is this basis orthogonal? orthonormal? neither?

B) Compute the coefficients of the vector $\boldsymbol{x} = 9\hat{\imath} - 5\hat{\jmath}$ in this $(\boldsymbol{u}, \boldsymbol{v})$ basis using matrices and vectors. Identify your final answer as $\boldsymbol{u}$- and $\boldsymbol{v}$-coordinates.

C) What are the coefficients of the vector $\boldsymbol{u} - \boldsymbol{v}$ in this $(\boldsymbol{u}, \boldsymbol{v})$ basis?

PROBLEM 8. Consider the linear transformation from $\mathbb{R}^4$ to $\mathbb{R}^4$ given by:

$$f\begin{pmatrix} x \\ y \\ u \\ v \end{pmatrix} = \begin{pmatrix} Cx + 2Cy - 2u \\ 3y + 4u - 3Cv \\ -u + Cv \\ 2u - 6v \end{pmatrix}$$

where C is an unknown constant.

A) What matrix A represents this linear transformation? That is, for what matrix A is $f(\boldsymbol{x}) = A\boldsymbol{x}$, where $\boldsymbol{x} = (x \;\; y \;\; u \;\; v)^T$?

B) For which value(s) of C is A invertible? Use the determinant to answer.

PROBLEM 9. Compute the determinants of the following matrices, using any method you wish.

A) $\begin{bmatrix} 0 & -9 & 5 \\ -1 & 2 & 17 \\ 0 & 3 & 0 \end{bmatrix}$ B) $\begin{bmatrix} 1 & 3 & 0 & 0 \\ -4 & -8 & 0 & 0 \\ 0 & 0 & 3 & 3 \\ 0 & 0 & 6 & 5 \end{bmatrix}$ C) $\begin{bmatrix} 2 & -4 & -9 & 15 \\ 0 & 1 & 2 & -18 \\ 4 & 0 & 4 & 7 \\ 1 & 0 & 1 & 3 \end{bmatrix}$

PROBLEM 10. Consider the following matrices:

$$A = \begin{bmatrix} -4 & 5 & 0 & 0 & 0 \\ 1 & -2 & 0 & 0 & 0 \\ 0 & 0 & 1 & 4 & 0 \\ 0 & 0 & 2 & 3 & 0 \\ 0 & 0 & 0 & 0 & -1 \end{bmatrix} \;:\; B = \begin{bmatrix} 7 & 6 & 0 & 0 & 0 \\ 6 & 5 & 0 & 0 & 0 \\ 0 & 0 & 2 & 8 & 0 \\ 0 & 0 & 1 & 5 & 0 \\ 0 & 0 & 0 & 0 & 3 \end{bmatrix}$$

Compute: A) $\det(AB)$; and B) $\det(B^{-1})$.

PROBLEM 11. Consider the following matrices:

$$A = \begin{bmatrix} 0 & 1 \\ 2 & 0 \\ 3 & -4 \end{bmatrix} \;:\; B = \begin{bmatrix} 3 & 2 & -9 \\ 0 & 0 & 4 \\ 6 & 3 & 17 \end{bmatrix} \;:\; C = \begin{bmatrix} -1 & 0 & 1 \\ 0 & 1 & 1 \\ 1 & 0 & 1 \end{bmatrix} \;:\; D = \begin{bmatrix} 3 & 0 & 5 \\ 2 & -1 & 0 \end{bmatrix}$$

Compute/simplify the following determinants, if possible: if not, explain why it's not possible.

A) $\det(B)$; B) $\det(C^3)$; C) $\det(CA)$; D) $\det(DA)$

PROBLEM 12. Consider the following matrices:

$$A = \begin{bmatrix} -1 & -1 \\ 2 & 0 \\ 3 & 1 \end{bmatrix} : B = \begin{bmatrix} -1 & 0 & 0 \\ 0 & 1 & -6 \\ 0 & -2 & 9 \end{bmatrix} : C = \begin{bmatrix} 3 & 0 & 0 \\ 0 & 2 & 5 \\ 0 & 4 & 8 \end{bmatrix} : D = \begin{bmatrix} 1 & 2 & -3 \\ 0 & 3 & 1 \end{bmatrix}$$

Compute/simplify the determinants of the following inverses, if possible: if not, explain why it's not possible.

A) $\det(C^{-1})$; B) $\det(BC)$; C) $\det(DA)$; D) $\det(BA)$

PROBLEM 13. Compute the determinants of the following matrices, using any method you wish.

A) $\begin{bmatrix} 1 & -5 & 3 \\ -2 & 19 & 7 \\ 0 & 2 & 0 \end{bmatrix}$ B) $\begin{bmatrix} 2 & 4 & 0 & 0 \\ 5 & 8 & 0 & 0 \\ 0 & 0 & -4 & 5 \\ 0 & 0 & 6 & -9 \end{bmatrix}$ C) $\begin{bmatrix} 2 & 4 & 7 & 19 & -9 \\ 0 & 5 & 5 & 12 & 8 \\ 0 & 0 & -1 & 0 & 7 \\ 0 & 0 & 0 & -5 & 7 \\ 0 & 0 & 0 & 1 & 3 \end{bmatrix}$

PROBLEM 14. Consider the following matrices A and B:

$$A = \begin{bmatrix} 1 & -2 & 4 & 3 \\ 2 & -1 & 0 & 5 \\ 0 & 0 & 1 & 3 \\ 0 & 0 & -2 & 1 \end{bmatrix} \qquad B = \begin{bmatrix} 1 & 2 & 0 & 0 \\ -2 & -1 & 0 & 0 \\ 4 & 0 & 1 & -2 \\ 3 & 5 & 3 & 1 \end{bmatrix}$$

Compute the determinant of the matrix AB.

PROBLEM 15. Consider the matrix A expressed as a product of two other matrices:

$$A=\begin{bmatrix}-1 & -4 & 2 & 3\\ 0 & 3 & 9 & -8\\ 0 & 0 & 2 & 7\\ 0 & 0 & 0 & 5\end{bmatrix}\begin{bmatrix}2 & 0 & 0 & 0\\ 13 & 1 & 0 & 0\\ 6 & 3 & -3 & 0\\ -17 & -7 & 4 & 1\end{bmatrix}$$

A) Explain: why is A an invertible matrix?

B) What is the determinant of A^{-1}?

PROBLEM 16. Use row reduction to compute the determinant of this matrix:

$$\begin{bmatrix}1 & -2 & 8 & 19 & -5 & 11\\ 3 & -1 & 3 & 5 & 10 & -9\\ 0 & 0 & 0 & 0 & -3 & 2\\ 0 & 0 & 0 & 3 & -5 & 17\\ 0 & 0 & 1 & -1 & 2 & 0\\ 0 & 0 & 0 & 0 & 0 & 3\end{bmatrix}$$

PROBLEM 17. Given the following facts about 3-by-3 matrices A and B:

1) $AB = BA$
2) $\det A^2 = 4$
3) $\det B^2 = 9$
4) $\det(A + B) = 5$
5) $\det(A - B) = -7$

Compute the following: A) $\det A$; B) $\det B$; C) $\det(A^2 - B^2)$

PROBLEM 18. Assume the following matrix has determinant equal to 7.

$$\begin{bmatrix}a_1 & a_2 & a_3\\ b_1 & b_2 & b_3\\ c_1 & c_2 & c_3\end{bmatrix}$$

Compute the determinants of the following matrices:

$$A=\begin{bmatrix}c_1 & c_2 & c_3\\ 2c_1+b_1 & 2c_2+b_2 & 2c_3+b_3\\ -3a_1 & -3a_2 & -3a_3\end{bmatrix} : B=\begin{bmatrix}-a_1 & b_1+c_1 & b_1\\ -a_2 & b_2+c_2 & b_2\\ -a_3 & b_3+c_3 & b_3\end{bmatrix}$$

$$C=\begin{bmatrix}a_1 & a_2 & 0 & a_3 & 0\\ 0 & 0 & -4 & 0 & 0\\ b_1 & b_2 & 0 & b_3 & 0\\ c_1 & c_2 & 0 & c_3 & 0\\ 0 & 0 & 0 & 0 & 3\end{bmatrix}$$

ANSWERS & HINTS

PROBLEM 1. A) orthogonal ; B) $\boldsymbol{v}=\frac{31}{26}\boldsymbol{a}+\frac{1}{13}\boldsymbol{b}$; C) (1,1)

PROBLEM 2. B) $CBA=\begin{bmatrix}1 & 2\\ 1 & 0\end{bmatrix}$; C) $\begin{pmatrix}1\\ 2\end{pmatrix}$

PROBLEM 3. A) ; B)

PROBLEM 4. A) $AB=\begin{bmatrix}4 & -2\\ 10 & 1\end{bmatrix}$; B) $B=\begin{bmatrix}1 & -7/2\\ 2 & 5\end{bmatrix}$

PROBLEM 5. B) $\begin{bmatrix}-1/2 & 0\\ 0 & 2\end{bmatrix}$

PROBLEM 6. A) $A=\begin{bmatrix}0 & -2 & 0 & 0\\ 2 & 0 & 0 & 0\\ 0 & 0 & -6 & 4\\ 0 & 0 & 3 & -2\end{bmatrix}$

B) nope ; C) counterclockwise rotation by $\pi/2$ and rescale by 2

PROBLEM 7. A) neither ; B) $\boldsymbol{x} = 17\boldsymbol{u} - 19\boldsymbol{v}$; C) $(1,-1)$

PROBLEM 8. A) $A = \begin{bmatrix} C & 2C & -2 & 0 \\ 0 & 3 & 4 & -3C \\ 0 & 0 & -1 & C \\ 0 & 0 & 2 & -6 \end{bmatrix}$; B) $C \neq 0, 3$

PROBLEM 9. A) -15 ; B) -12 ; C) 15

PROBLEM 10. A) -90 ; B) $-1/6$

PROBLEM 11. A) 12 ; B) -8 ; C) nope ; D) -4

PROBLEM 12. A) $-1/12$; B) -36 ; C) 30 ; D) nope

PROBLEM 13. A) -26 ; B) -24 ; C) 220

PROBLEM 14. $\det(AB) = (\det A)(\det B) = (21)(21) = 441$

PROBLEM 15. A) $\det A = (-30)(-6) = 180 \neq 0$; B) $\det(A^{-1}) = (\det A)^{-1} = 1/180$

PROBLEM 16. $\det = 45$

PROBLEM 17. A) 2 ; B) 3 ; C) -35

PROBLEM 18. A) $\det A = 21$; B) $\det B = -7$; C) $\det C = 84$

VOLUME II : DERIVATIVES

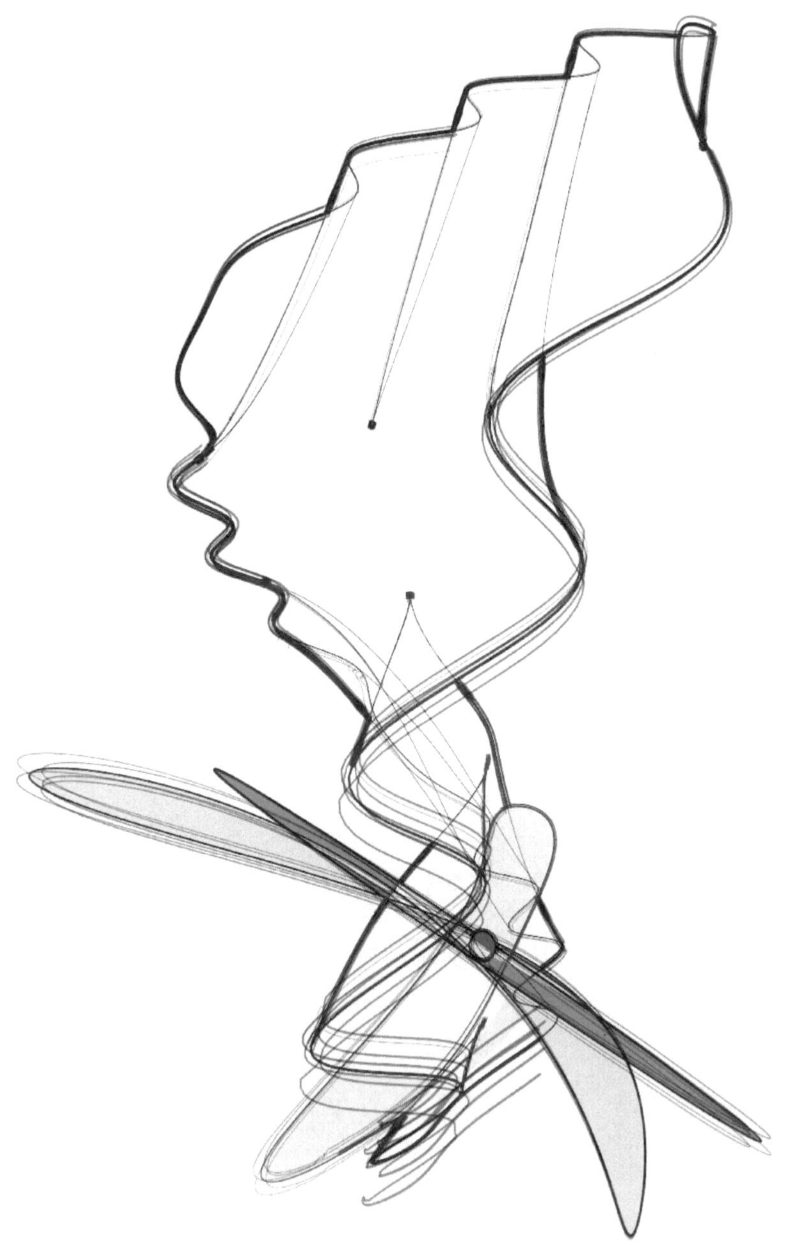

Week 5 :
The Derivative

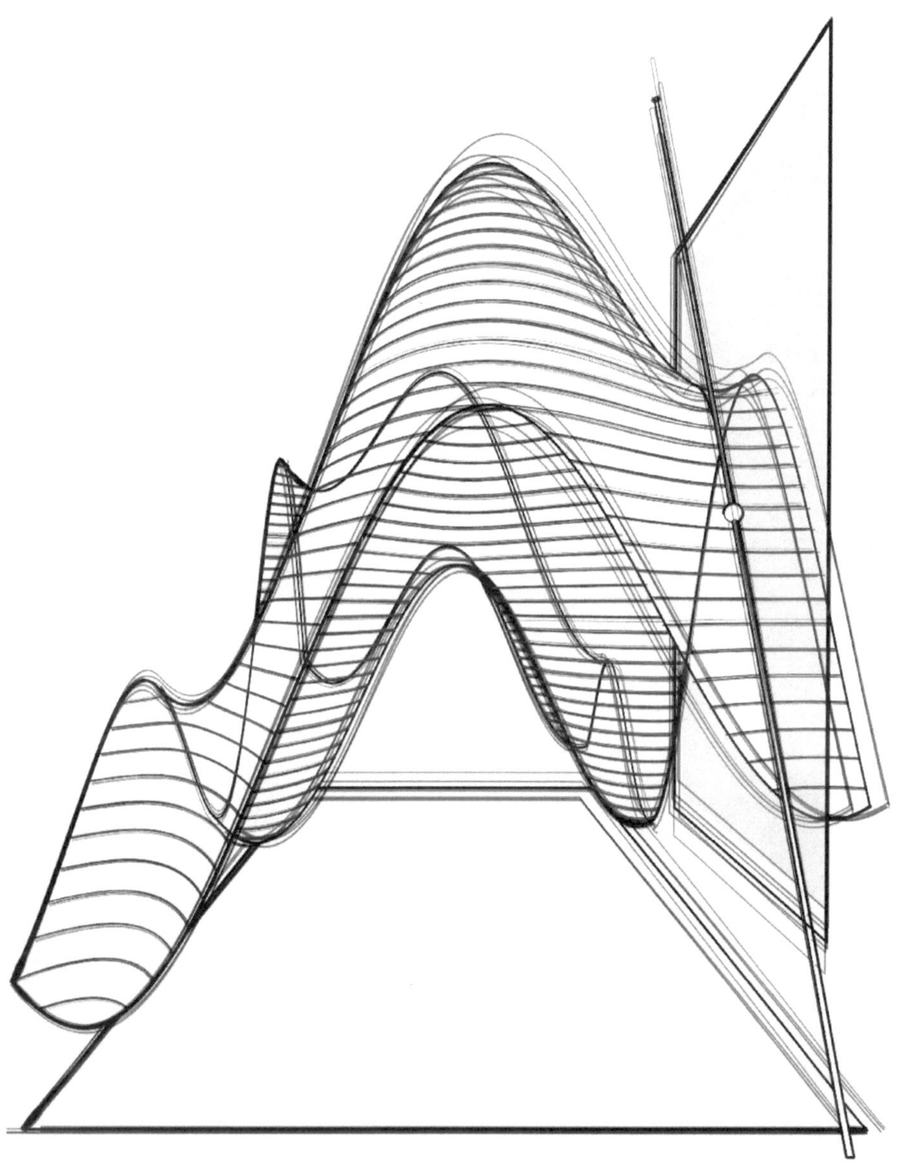

OUTLINE

MATERIALS: Calculus BLUE : Vol 2 : Chapters 1-4

TOPICS:

- ▷ Multivariate functions and their applications
- ▷ Partial derivatives: computation and interpretation
- ▷ The derivative as a matrix of partials
- ▷ The derivative as a linear transformation on vectors of rates of change
- ▷ Sensitivity of input-output pairs based on partial derivatives
- ▷ Definition of the derivative
- ▷ The derivative of the polar coordinate transformation
- ▷ Continuous but non-differentiable functions
- ▷ The derivative as a 1st order term in a Taylor expansion

LEARNING OBJECTIVES:

- ▷ Manipulate functions having multiple inputs and outputs
- ▷ Compute the partial derivatives of a function
- ▷ Compute the derivative as a matrix
- ▷ Evaluate the derivative at different inputs
- ▷ Discern number of inputs and outputs of a function based on derivative
- ▷ Use the derivative to transform vectors of rates of change of inputs
- ▷ Use the derivative to discern sensitivities of inputs/outputs

PRIMER

This is the first week in which we can be said to be doing calculus with truly multivariate functions with multiple inputs and multiple outputs. Our work with matrices as linear transformations has prepared us for this moment, yielding a plentiful source of functions $A: \mathbb{R}^n \to \mathbb{R}^m$ of a simple (linear) nature.

PARTIAL DERIVATIVES. For a function $f: \mathbb{R}^n \to \mathbb{R}^m$ of n variables $\boldsymbol{x} \in \mathbb{R}^n$, we can consider what happens to the m outputs $f_1, f_2, \dots, f_m$ if only one of the inputs, x_j, is varied. If we restrict attention to the ith output f_i, keeping all but one input variable x_j fixed, then f is in the familiar form from single-variable calculus. The derivative of the ith output f_i with respect to the jth input x_j is the *partial derivative* $\partial f_i / \partial x_j$. The differentiation operator with respect to x_j holding all other inputs constant is denoted $\partial / \partial x_j$.

The collection of all partial derivatives of a function f can be unwieldy, depending on the number of inputs and outputs. Our strategy for managing all this data is, at first, to use a matrix as a convenient data structure. This matrix, often called the *Jacobian*, will henceforth be called simply *the derivative.* It is denoted with square brackets to reinforce its matrix-like nature:

$$[Df] = \left[\frac{\partial f_i}{\partial x_j}\right]_{ij}$$

It is important to remember that the columns of the derivative correspond to the inputs of f and the rows correspond to the outputs of f: a little practice at this will pay off in this and future weeks.

As in single variable calculus, there is an important difference between the derivative evaluated at a particular point and the un-evaluated derivative, whose entries are functions of the input variables. The derivative evaluated at a point is a matrix whose entries are numerical. These numbers – the partial derivatives – can be interpreted as *sensitivities* of input-output pairs: the sensitivity of the ith output with respect to the jth input. It is not itself a rate of change or an actual change in output. Like a slope, the partial derivative is a ratio of output-to-input rates of change, and, like slope, the sign matters greatly, indicating positive or negative correlation between input and output. As always, changing the evaluation point of the derivative can change the entries dramatically (but hopefully continuously).

THE DERIVATIVE. Why is the derivative best presented as a matrix? Is this simply a convenient data structure or is there something deeper at work? The derivative is more than a matrix: *it is a linear transformation*. When evaluated at a fixed input point $\boldsymbol{a}$, the derivative is a linear transformation $[Df]_{\boldsymbol{a}}$ which takes a vector of rates of change of inputs to f at $\boldsymbol{a}$ to a vector of rates of change of the outputs of f at the point $f(\boldsymbol{a})$. This is the first deep idea in multivariable calculus, and it takes some effort to grasp.

Thinking in terms of a parametrized curve or surface can be helpful (*cf.* Week 2 and the discussion problems below), as is working with a simple example, such as polar coordinates. The true power of the derivative as a linear transformation lies in its ability to handle very large complex sets of inputs and outputs. Given a derivative at a point, call this matrix $A = [Df]_{\boldsymbol{a}}$, and let $\boldsymbol{h}$ be a vector of rates of change of inputs. The vector of rates of change of outputs is easily computed as the product $\boldsymbol{b} = A\boldsymbol{h}$. If, on the other hand, a vector $\boldsymbol{b}$ of rates of change of outputs is desired, with the goal being to find the precise rates of change of inputs that effect this, then we are again faced with a linear system: solve $A\boldsymbol{h} = \boldsymbol{b}$ for $\boldsymbol{h}$. All the possible complexities – does a solution exist and is it unique – are a solved problem, thanks to the work done in Weeks 3-4.

One may wonder what other results from matrix algebra will be useful to us in light of derivatives as linear transformations. What does matrix multiplication or the determinant mean? We will return to this question next week.

THE DEFINITION. The derivative $[Df]$ has thus far been defined solely in terms of (computable and interpretable) partial derivatives. A more formal definition is necessary and proper. The derivative of f at a point $\boldsymbol{a}$ is the linear transformation $[Df]_{\boldsymbol{a}}$ which acts on vectors $\boldsymbol{h}$ such that

$$\lim_{|\boldsymbol{h}|\to 0^+} \frac{f(\boldsymbol{a}+\boldsymbol{h}) - f(\boldsymbol{a}) - [Df]_{\boldsymbol{a}}\boldsymbol{h}}{|\boldsymbol{h}|} = \boldsymbol{0},$$

where the right hand side is the zero vector. This is a limit not as the vector $\boldsymbol{h}$ goes to zero, but rather as the length of $\boldsymbol{h}$ goes to zero. We cannot use the "usual" definition of the derivative since we cannot divide by a vector. Details of the definition are not essential to our story, though proper definitions are the glory of Mathematics. If it seems confusing now, it may be a good idea to revisit this after learning about Taylor expansion in Week 7.

One way in which this more formal definition is helpful appears when working with functions whose partial derivatives are simply too numerous to handle. For example, the function S which takes an n-by-n matrix A to its square $S(A) = A^2$ can be thought of as a function $S: \mathbb{R}^{n^2} \to \mathbb{R}^{n^2}$ whose partial derivatives are not readily computed except in the case of small values of n.

Nevertheless, by stacking the (rates of) changes to inputs as a matrix H, one has via a Taylor-expansion-type argument that

$$S(A+H)=(A+H)^2=S(A)+AH+HA+H^2$$

Thus, by the formal definition, $[DS]_A H = AH+HA$: *cf.* the single-variable calculus result that $(x^2)'|_a = 2a$.

DISCUSSION

QUESTION 1. Compute the following partial derivatives.

$$F=\frac{x^2\sqrt{y^3}}{z^4} \quad \Rightarrow \quad \frac{\partial F}{\partial x}=\cdots \quad : \quad \frac{\partial F}{\partial y}=\cdots \quad : \quad \frac{\partial F}{\partial z}=\cdots$$

Start off by computing enough examples that students feel comfortable with partial derivatives. This will not take long. It's good to say out loud the first few times "Ok, if everything but x is a constant..."

QUESTION 2. Recall from Week 1 working with quadratic surfaces. Compute the derivative – as a matrix – of the following parametrization of a paraboloid:

$$z=x^2+y^2 \quad \Rightarrow \quad G\begin{pmatrix}s\\t\end{pmatrix}=\begin{pmatrix}s\\t\\s^2+t^2\end{pmatrix}$$

This is simple – for now – but will be important in understanding why $[DG]$ is more than a matrix. In this case (where the inputs are not x and y), does it matter what order we use for the inputs? What determines which input is first? This is a good time to emphasize writing the inputs and outputs as ordered lists of variables. If you want to permute, you can, but consistency is vital.

QUESTION 3. Thinking in terms of our preparatory work in Week 2 may be helpful. Explain the velocity vector of a parametrized curve $\gamma(t)$ in terms of the definition of a derivative that we have learned this week.

Emphasize the relationship between the velocity vector γ' and the derivative $[D\gamma]$, both of which are column vectors. Students with a physics background may find it helpful to think of the velocity vector as a linear transformation – give me a Δt and get a $\Delta\mathbf{x}$. Be sure to remind students that the derivative can be very different (or even zero) depending on where it is evaluated, just as in single variable calculus.

QUESTION 4. Why is it that a derivative is a linear transformation? Consider a parametrized surface in 3-D, of the form $f:\mathbb{R}^2\to\mathbb{R}^3$. (Question 2 may still be on the board...) Assume input parameters s and t with outputs x, y, z each depending on the two parameters. What happens if, at a particular input (s_0, t_0) we fix one of the parameters and let the other change? What does the output look like?

This is a good excuse to draw a picture of a surface in 3-D on the board. Using two colors for the image of the s axis and the t axis is helpful. Try to get students to think in terms of velocity vectors. Once the axes – the partial derivatives – are grasped, then ask what happens if both inputs are increased? What if one is increased and the other decreased? Be sure to recall the notion of a basis from Week 4. We will revisit this idea of a tangent plane in Week 7.

QUESTION 5. Consider the linear function

$$\begin{pmatrix}u\\v\end{pmatrix}=\begin{pmatrix}x+2y\\3x+5y\end{pmatrix}=\begin{bmatrix}1&2\\3&5\end{bmatrix}\begin{pmatrix}x\\y\end{pmatrix}$$

Compute all the partials $\partial v/\partial x$, etc. Then, invert the transformation and compute the partials of the inverses. Is $\partial x/\partial v$ equal to the reciprocal of $\partial v/\partial x$? Compare with the single-variable case of $v(x) = 3x$. What is different in this case?

This foreshadows the Chain Rule and Inverse Function Theorem of Week 6, but, more importantly, this stresses the need to understand the entire derivative, and not simply keep a list of partials.

QUESTION 6. Think back to single-variable calculus... What is the derivative of the affine (linear + constant) function $y = ax + b$? Of course, the derivative is a constant. Now, what is the derivative of the vector-valued function $\mathbf{y} = A\mathbf{x} + \mathbf{b}$?

Clever students will guess at the correct answer. It is worth doing an explicit example with, say, a randomly-generated matrix A of size 2-by-3, just to emphasize the conventions of rows/columns being what they are. This problem also foreshadows the Taylor perspective of Week 7.

QUESTION 7. The following *Cobb-Douglas* model is a classic in basic Economics. One models production, P, as a function of materials, M, and labor, L, via

$$P = \kappa M^{\alpha} L^{\beta},$$

where $\kappa > 0$ and $0 < \alpha, \beta < 1$ are constants and $\alpha + \beta = 1$. If the investment in labor is increased and the investment in materials is decreased at an equal rate, what is the impact on production?

One may wish to begin with a discussion of the model: why these fractional powers? The crucial idea is that if one doubles M and L, one should double P (think: clone the factory). This is not important to this week's material, but it is a good review of dimensional-analytic thinking.

In this – and perhaps other problems – students who have seen some multivariable calculus before will want to think of everything as a function of time, t, and then use Chain Rule arguments to get a rate of change. This is a good exercise: do it both ways and see how the matrix version compares to the old-fashioned version. The matrix approach seems less intuitive and more cumbersome, to be fair. However – and this is crucial – ask students to consider what happens in the case of a model that is not a textbook cartoon. What happens when there are a hundred products being built by a dozen overlapping teams of workers managing a supply chain of a thousand different components? How should investments be allocated among the different labor teams and in which input material streams? How does one measure rates of change of productivity in the case of a hundred outputs? Not everything can be reduced to a function of a single (time) parameter.

QUESTION 8. The following function gives the midpoint deflection u of a beam of length L supported at the endpoints, with a cross section of width w and height h, bearing a load of weight F:

$$u = \frac{FL^3}{4Ewh^2}$$

Here E is a constant that depends on the material from which the beam is made. Compute the derivative $[Du]$ and evaluate at $L = 4, F = 100, E = 1, w = 3, h = 2$. At these values, what happens to the deflection if each input is increased at a unit rate? Does the beam deflect more or less?

This problem is perhaps too much in-the-weeds for non-engineering students, but this function has more than two inputs and one cannot automatically guess the correct answer.

QUESTION 9. Consider a function f such that, at a particular point,

$$[Df]\begin{pmatrix}1\\-1\end{pmatrix} = \begin{pmatrix}3\\-2\end{pmatrix}$$

Start off by asking how many inputs and outputs f has.

What happens if inputs change at rates $\boldsymbol{h} = (-2,2)^T$? *This is doable.*
What if $\boldsymbol{h} = (3,3)^T$? *This is not doable, but why not?*

What if you also know that $[Df]\begin{pmatrix}1\\2\end{pmatrix} = \begin{pmatrix}4\\-4\end{pmatrix}$? Can you answer the previous?

QUESTION 10. The derivative of a function f evaluated at an input $\boldsymbol{a}$ equals

$$[Df]_{\boldsymbol{a}} = \begin{bmatrix} -1 & 5 & -3 \\ 0 & 1 & 4 \\ 1 & -1 & 7 \\ 2 & 0 & -6 \end{bmatrix}$$

How many inputs and outputs does f have? Which input-output pair has the greatest sensitivity at this point? If all the inputs are decreasing at the same rate, which outputs are increasing? If the first and second inputs are increasing at the same rate, with the third input unchanged, which output is most sensitive to the change?

There are many similar questions one can ask with a randomly generated matrix.

QUESTION 11. The example of a function acting on square matrices from the videotext was... intimidating. But let's keep going! What is the derivative of the functions on square matrices given by $f(A) = A^3$ or even $f(A) = A^{-1}$ (assuming invertibility)? Recall, you do not want to compute partials here!

This is intimidating, but it emphasizes the definition of the derivative, and it gets at the Taylor expansion approach. The cube function seems easy – this is just the Binomial Theorem, right?

$$(A+H)^3 = A^3 + 3A^2H + 3AH^2 + H^3$$

Oops... What have we forgotten here? AH-HA! Commutativity should not be assumed...

The inverse is harder: use the geometric series to get the 1st-order term as $-A^{-1}HA^{-1}$, since

$$(A+H)^{-1} = \left(A\left(I - (-A^{-1}H)\right)\right)^{-1} = \left(I - (-A^{-1}H)\right)^{-1}A^{-1} = \left(I - A^{-1}H + O(H^2)\right)A^{-1}$$
$$= A^{-1} - A^{-1}HA^{-1} + O(H^2)$$

What elementary single-variable calculus result does this remind you of?

ASSESSMENT PROBLEMS

PROBLEM 1. Consider the following function:

$$f\begin{pmatrix}u\\v\\w\end{pmatrix} = \begin{pmatrix}u^2v^{-3}w\\2u-5w\\uv-vw\end{pmatrix}$$

A) Compute the derivative $[Df]$, showing work.

B) Evaluate this derivative at the point where $u = 1, v = -1, w = 2$.

C) If, at this point (using your result from part (B)), all the inputs are *decreasing* at the same rate, which output is increasing the most?

PROBLEM 2. At a particular point, a function f has derivative

$$[Df]_a = \begin{bmatrix} 1 & 0 & -1 \\ 4 & -8 & 2 \\ 2 & 13 & -1 \\ -1 & 7 & 0 \\ 3 & 0 & -3 \end{bmatrix}$$

A) How many inputs and outputs does f have?

B) Assume, at this point, the first input is increasing at a unit rate; the last input is decreasing at twice this unit rate; and all other inputs are not changing. Then, at what rates are the outputs changing?

C) If, at this point, the inputs are changing at a rate such that the last output is not changing (*i.e.*, the rate of change of the last output is zero), then what can you say about the rates of change of the inputs?

PROBLEM 3. Consider the following function:

$$f\begin{pmatrix} x \\ y \\ s \\ t \end{pmatrix} = \begin{pmatrix} xy^2 + 5t \\ xs - yt \\ 2x - 3y + s^2 t \end{pmatrix}$$

A) Compute the derivative $[Df]$.

B) Evaluate this derivative at the point where $x = 0, y = -1, s = 1, t = 2$.

C) If, at this point (using your result from part (B)), the s and t inputs are increasing at a unit rate and the x and y inputs are decreasing at the same unit rate, which output is changing the least?

PROBLEM 4. Consider the following function:

$$f\begin{pmatrix} x \\ y \\ z \end{pmatrix} = \begin{pmatrix} (1 + 2x + 3y)^{-1} \\ e^{2y-5z} \\ (x+2)(z-4) \end{pmatrix}$$

A) Compute the derivative $[Df]$, showing work.

B) Evaluate this derivative at the origin.

C) At the origin, which output is most sensitive with respect to which input? (That is, which input-output pair experiences the largest changes, when all other inputs are held constant.)

PROBLEM 5. Consider the following square matrix, depending on variables x and y:

$$A = \begin{bmatrix} x & 1 & 7x \\ 0 & 2 & y \\ 0 & x & 3y \end{bmatrix}$$

A) Compute and simplify the determinant $\det(A)$, showing work.

B) Define "sum(A)" to be the sum of all (nine) entries of the matrix A. Compute and simplify this sum(A), showing work.

C) Write out explicitly the function $f: \mathbb{R}^2 \to \mathbb{R}^2$ defined by

$$f\begin{pmatrix} x \\ y \end{pmatrix} = \begin{pmatrix} \det(A) \\ \text{sum}(A) \end{pmatrix},$$

using your results from above. Compute the derivative $[Df]$, showing work.

PROBLEM 6. Consider the three functions

$$f\begin{pmatrix}x\\y\\z\end{pmatrix}=\begin{pmatrix}(x-y)^3\\(z-1)(x+5)\end{pmatrix}\quad:\quad g\begin{pmatrix}s\\t\end{pmatrix}=\begin{pmatrix}5s-t\\\sin(2t-3s)\\e^{-s}-e^{-2t}\end{pmatrix}\quad:\quad h=f\circ g$$

The third, $h=f\circ g$, is the composition of f with g.

A) Compute the derivatives of f and g, evaluated at the origin.

B) Which of the following compositions are legal?

$$f\circ h\quad:\quad g\circ h\quad:\quad h\circ f\quad:\quad h\circ g$$

C) Are there any inputs at which the derivative of g vanishes?

PROBLEM 7. Consider the following function:

$$f\begin{pmatrix}x\\y\\z\end{pmatrix}=\begin{pmatrix}2z-\sqrt{x^2+y^2}\\y-\ln(z^2+x^2)\end{pmatrix}$$

A) Compute the derivative $[Df]$, showing work.

B) Evaluate this derivative at the point where $x=2, y=-3, z=0$.

C) If, at this point (using your result from part (B)), all the inputs are *decreasing* at the same rate, which output is increasing the most?

PROBLEM 8. Consider the following function:

$$\begin{pmatrix}y_1\\y_2\\y_3\\y_4\end{pmatrix}=f\begin{pmatrix}x_1\\x_2\\x_3\\x_4\end{pmatrix}=\begin{pmatrix}x_1^2-x_4^3\\x_1x_2x_3\\x_1+2x_2+4x_4\\x_1-x_4-x_2x_3\end{pmatrix}$$

A) Compute the derivative $[Df]$.

B) Which partial derivatives of f vanish (are equal to zero)? Please list.

C) Evaluate $[Df]$ at the origin: how many entries of this matrix are nonzero?

PROBLEM 9. Consider the following function:

$$f\begin{pmatrix}u\\v\\w\end{pmatrix}=\begin{pmatrix}e^w-2u\\u+v+w\\u^2w\\uv^2-3w\end{pmatrix}$$

A) Compute the derivative $[Df]$, showing work.

B) Evaluate this derivative at the point where $u=1, v=2, w=0$.

C) If, at this point (using your result from part (B)), all the inputs are *increasing* at a unit rate, which output is increasing the most?

PROBLEM 10. Assume that a function f has derivative

$$[Df]=\begin{bmatrix}2y&2x\\-3&5y\end{bmatrix}$$

A) How many inputs and outputs does f have?

B) Evaluate this derivative at the point where $x=2$ and $y=-1$.

C) Assume, at some point (x,y), the inputs are changing at rates $+2$ and -1 and the **outputs** are changing at rates -2 and $+4$. Write out an equation that expresses this, using vectors of rates of change and the derivative $[Df]$ as a linear transformation.

D) Using the equations from (C) above, solve for which point this is happening at.

PROBLEM 11. At a particular point, a function f has derivative

$$[Df]_a = \begin{bmatrix} 1 & -2 & 3 \\ 3 & 7 & 5 \\ -2 & -6 & -1 \\ 4 & -5 & 3 \\ 7 & 0 & -6 \end{bmatrix}$$

A) How many inputs and outputs does f have?

B) Assume that at this point you have to change each input independently, either increasing or decreasing it at a fixed rate. If your goal is to increase all the outputs, how would you toggle the changes to the inputs?

PROBLEM 12. Assume that at a particular input $\boldsymbol{a}$, a function f has derivative

$$[Df]_a = \begin{bmatrix} 1 & 2 & -1 & -5 & -3 \\ 2 & -7 & -2 & 1 & 0 \\ -3 & 4 & 2 & 3 & 8 \\ 0 & 2 & 1 & -2 & 0 \\ 3 & 0 & 6 & 7 & 2 \\ 0 & -7 & 0 & 0 & -1 \end{bmatrix}$$

A) How many inputs and outputs does f have?

B) If the first three inputs are increasing at a unit rate and the remaining inputs are decreasing at twice this rate; then at what rate is the fourth output changing? Be sure to explain and show your work.

C) If all the inputs are decreasing at a unit rate, which of the outputs is increasing the most?

PROBLEM 13. Assume that a function f has derivative at a point $\boldsymbol{a}$ given by

$$[Df]_a = \begin{bmatrix} 2 & 0 & 0 \\ 0 & 6 & -3 \\ 0 & 3 & -1 \end{bmatrix}$$

A) If, at this point, the three inputs are changing at rates $+4, -3$, and $+2$ respectively; then, at what rates are the three outputs changing?

B) If, at this point, the three inputs are changing at unknown rates c_1, c_2, and c_3 and the three outputs are changing at rates $-6, -3$, and $+1$ respectively; then, at what rates are the inputs changing?

PROBLEM 14. Consider the parametrized surface in 3-D given by

$$F\begin{pmatrix} s \\ t \end{pmatrix} = \begin{pmatrix} s^3 - 2t^2 + 7 \\ (s-1)(t-2) \\ s^2 - t^2 \end{pmatrix} \begin{matrix} \leftarrow x \\ \leftarrow y \\ \leftarrow z \end{matrix}$$

A) Compute the derivative of F.

B) Evaluate the derivative of F at $s = 1$ and $t = 2$.

C) Assume that you start at the point on the surface in which $s = 1$ and $t = 2$. Is it possible to increase both the s and t parameters at nonzero rates so that both the y- and z-components on the surface change at rate zero? If so, at what rates should these inputs be increased?

PROBLEM 15. Assume that at a point $\boldsymbol{a}$, the derivative of a function f equals

$$[Df]_a = \begin{bmatrix} 2 & -1 & 0 & 0 \\ -3 & 1 & 0 & 0 \\ 0 & 0 & 3 & -2 \\ 0 & 0 & -5 & 4 \end{bmatrix}$$

A) Which input-output pair has the highest sensitivity here? That is, which output varies the most with respect to which input?

B) If the 1st input is changing at rate +1; the 2nd at rate +2; the 3rd at +3; and the 4th at +4, then at what rates are the outputs changing?

C) If the outputs are all changing at a rate of +1, at what rates are the inputs changing?

PROBLEM 16. Consider linear transformations f and g whose matrices are given by:

$$f : \begin{bmatrix} 1 & 3 & 0 & -2 \\ -1 & 0 & 1 & 0 \\ 7 & 0 & 1 & 5 \end{bmatrix} \qquad g : \begin{bmatrix} 2 & 0 & 1 \\ 0 & 2 & -5 \\ 4 & -1 & 0 \\ -1 & 0 & 2 \end{bmatrix}$$

A) How many inputs and outputs does f have?

B) Which input-output pair of g is most sensitive to change?

C) Compute the derivative of the composition $f \circ g$ at the origin.

ANSWERS & HINTS

PROBLEM 1. A/B) $\begin{bmatrix} 2uv^{-3}w & -3u^2v^{-4}w & u^2v^{-3} \\ 2 & 0 & -5 \\ v & u-w & -v \end{bmatrix} \Rightarrow \begin{bmatrix} -4 & -6 & 1 \\ 2 & 0 & -5 \\ -1 & -1 & 1 \end{bmatrix}$; C) 1st output

PROBLEM 2. A) 3 inputs, 5 outputs ; B) $[Df]_a \begin{pmatrix} 1 \\ 0 \\ -2 \end{pmatrix} = \begin{pmatrix} 3 \\ 0 \\ 4 \\ -1 \\ 9 \end{pmatrix}$; C) the rates of change of the 1st and 3rd inputs are equal

PROBLEM 3. A/B) $\begin{bmatrix} y^2 & 2xy & 0 & 5 \\ s & -t & x & -y \\ 2 & -3 & 2st & s^2 \end{bmatrix} \Rightarrow \begin{bmatrix} 1 & 0 & 0 & 5 \\ 1 & -2 & 0 & 1 \\ 2 & -3 & 4 & 1 \end{bmatrix}$; C) 2nd output

PROBLEM 4. A/B) $\begin{bmatrix} 2(1+2x+3y)^{-2} & 3(1+2x+3y)^{-2} & 0 \\ 0 & 2e^{2y-5z} & -5e^{2y-5z} \\ z-4 & 0 & x+2 \end{bmatrix} \Rightarrow \begin{bmatrix} 2 & 3 & 0 \\ 0 & 2 & -5 \\ -4 & 0 & 2 \end{bmatrix}$;
C) the 2nd output and 3rd input

PROBLEM 5. $f\begin{pmatrix} x \\ y \end{pmatrix} = \begin{pmatrix} 5xy \\ 9x+4y+3 \end{pmatrix}$; $[Df] = \begin{bmatrix} 5y & 5x \\ 9 & 4 \end{bmatrix}$

PROBLEM 6. A) $[Df] = \begin{bmatrix} 0 & 0 & 0 \\ -1 & 0 & 5 \end{bmatrix}$, $[Dg] = \begin{bmatrix} 5 & -1 \\ -3 & 2 \\ -1 & 2 \end{bmatrix}$; B) $g \circ h$; C) no

PROBLEM 7. A) $[Df] = \begin{bmatrix} -x(x^2+y^2)^{-\frac{1}{2}} & -y(x^2+y^2)^{-\frac{1}{2}} & 0 \\ -2x(x^2+z^2)^{-1} & 1 & -2z(x^2+z^2)^{-1} \end{bmatrix}$;

B) $\begin{bmatrix} -2/\sqrt{13} & 3/\sqrt{13} & 2 \\ -1 & 1 & 0 \end{bmatrix}$; C) $\begin{pmatrix} -2-1/\sqrt{13} \\ 0 \end{pmatrix}$

PROBLEM 8. A) $[Df] = \begin{bmatrix} 2x_1 & 0 & 0 & -3x_4 \\ x_2x_3 & x_1x_3 & x_1x_2 & 0 \\ 1 & 2 & 0 & 4 \\ 1 & -x_3 & -x_2 & -1 \end{bmatrix}$; C) $\begin{bmatrix} 0 & 0 & 0 & 0 \\ 0 & 0 & 0 & 0 \\ 1 & 2 & 0 & 4 \\ 1 & 0 & 0 & -1 \end{bmatrix}$

PROBLEM 9. A) $[Df] = \begin{bmatrix} -2 & 0 & e^w \\ 1 & 1 & 1 \\ 2uw & 0 & u^2 \\ v^2 & 2uv & -3 \end{bmatrix}$; B) $\begin{bmatrix} -2 & 0 & 1 \\ 1 & 1 & 1 \\ 0 & 0 & 1 \\ 4 & 4 & -3 \end{bmatrix}$; C) 4th output

PROBLEM 10. A) 2-by-2 ; B) $\begin{bmatrix} -2 & 4 \\ -3 & -5 \end{bmatrix}$; C) $\begin{bmatrix} 2y & 2x \\ -3 & 5y \end{bmatrix}\begin{pmatrix} 2 \\ -1 \end{pmatrix} = \begin{pmatrix} -2 \\ 4 \end{pmatrix}$; D) $\begin{pmatrix} x \\ y \end{pmatrix} = \begin{pmatrix} -3 \\ -2 \end{pmatrix}$

PROBLEM 11. A) 3 inputs, 5 outputs ; B) increase; decrease; increase

PROBLEM 12. A) 5 inputs, 6 outputs ; B) 7 ; C) 6th output

PROBLEM 13. A) $\begin{pmatrix} 8 \\ -24 \\ -11 \end{pmatrix}$; B) $\begin{pmatrix} c_1 \\ c_2 \\ c_3 \end{pmatrix} = \begin{bmatrix} 1/2 & 0 & 0 \\ 0 & -1/3 & 1 \\ 0 & -1 & 2 \end{bmatrix}\begin{pmatrix} -6 \\ -3 \\ 1 \end{pmatrix} = \begin{pmatrix} -3 \\ 2 \\ 5 \end{pmatrix}$

PROBLEM 14. A/B) $[Df] = \begin{bmatrix} 3s^2 & -4t \\ t-2 & s-1 \\ 2s & -2t \end{bmatrix} \Rightarrow \begin{bmatrix} 3 & -8 \\ 0 & 0 \\ 2 & -4 \end{bmatrix}$; B) $\begin{pmatrix} h_s \\ h_t \end{pmatrix} = C\begin{pmatrix} 2 \\ 1 \end{pmatrix}$

PROBLEM 15. A) 3rd input, 4th output ; B) $[Df]_a \begin{pmatrix} 1 \\ 2 \\ 3 \\ 4 \end{pmatrix} = \begin{pmatrix} 0 \\ -1 \\ 1 \\ 1 \end{pmatrix}$; C) $\begin{pmatrix} -2 \\ -5 \\ 3 \\ 4 \end{pmatrix}$

PROBLEM 16. A) 4 inputs, 3 outputs; B) 2nd output, 3rd input ; C) as these are linear transformations, composition is matrix multiplication and

$$[D(f \circ g)] = \begin{bmatrix} 1 & 3 & 0 & -2 \\ -1 & 0 & 1 & 0 \\ 7 & 0 & 1 & 5 \end{bmatrix}\begin{bmatrix} 2 & 0 & 1 \\ 0 & 2 & -5 \\ 4 & -1 & 0 \\ -1 & 0 & 2 \end{bmatrix} = \begin{bmatrix} 4 & 6 & -18 \\ -6 & -1 & -1 \\ 14 & -1 & 17 \end{bmatrix}$$

Week 6 : Differentiation

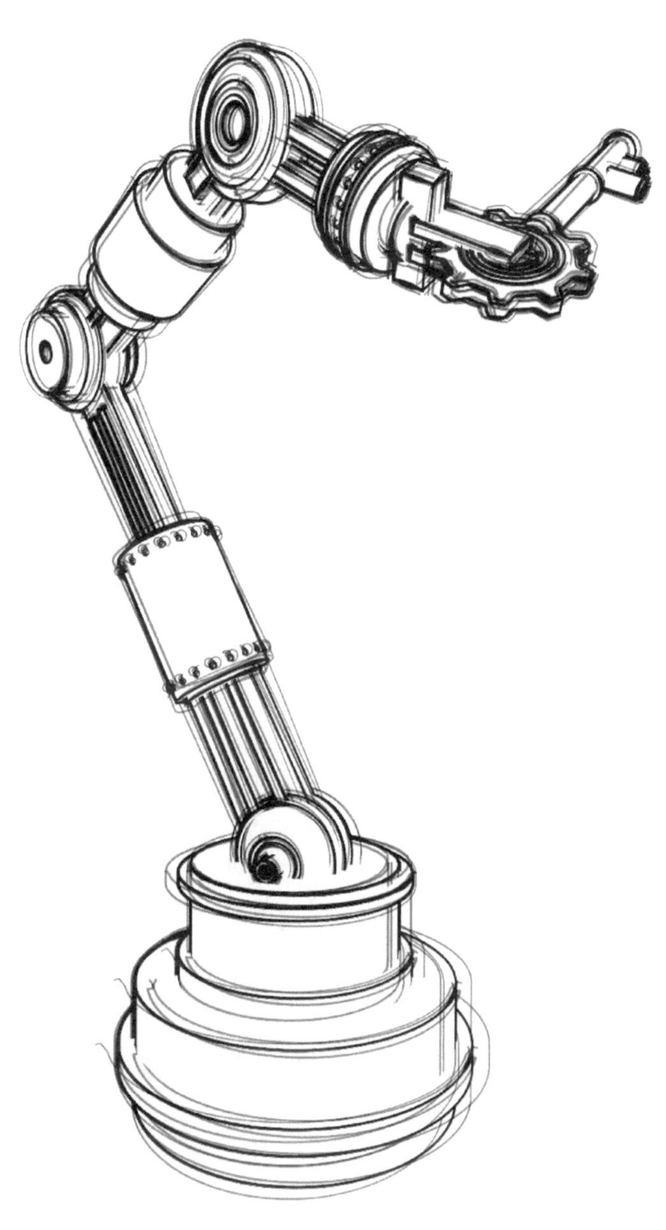

OUTLINE

MATERIALS: Calculus BLUE : Vol 2 : Chapters 5-8

TOPICS:

- ▷ Differentiation as a linear operator
- ▷ The Chain Rule and composition
- ▷ Applications of the Chain Rule
- ▷ BONUS : the material derivative
- ▷ Inverse functions in multivariate setting
- ▷ The Inverse Rule for derivatives
- ▷ The Inverse Function Theorem and its uses
- ▷ BONUS : The Implicit Function Theorem and its uses

LEARNING OBJECTIVES:

- ▷ Use linearity to compute derivatives of linear combinations of functions
- ▷ Infer when functions can and cannot be composed
- ▷ Use the Chain Rule to compute derivatives of compositions
- ▷ Explain the idea of an inverse of a multivariate function
- ▷ Explain the difference between local and global invertibility
- ▷ Use the Inverse Function Theorem to determine local invertibility
- ▷ Use the Inverse Rule to determine the derivative of an inverse

PRIMER

Recall from single-variable calculus the array of derivative rules which make computing derivatives of very complicated functions relatively procedural. This week, we recapitulate those rules and push out to new and deeper results.

THE RULES. The primal differentiation rule is that of linearity: differentiation is a *linear operator*. This is perhaps remembered from single variable calculus as the *Addition Rule*, though linearity also entails a scalar-multiplication rule as well. For any scalar c and compatibly-sized functions f and g, $[D(f+cg)] = [Df] + c[Dg]$.

The most useful differentiation rule is the *Chain Rule* governing the composition of functions. Recall from single variable calculus that, given functions g and f, the derivative of the composition $f \circ g$ is the product of the derivatives, evaluated at the correct points: $(f \circ g)'_a = f'_{g(a)} g'_a$. (If not immediately recalled, take the derivative of $\sin x^2$ to see what the hand remembers.) The Chain Rule for multivariate functions is the same. For compatibly sized functions $g\colon \mathbb{R}^n \to \mathbb{R}^p$ and $f\colon \mathbb{R}^p \to \mathbb{R}^m$ the composition $(f \circ g)\colon \mathbb{R}^n \to \mathbb{R}^m$ has derivative

$$[D(f \circ g)]_{\boldsymbol{a}} = [Df]_{g(\boldsymbol{a})}[Dg]_{\boldsymbol{a}}\,.$$

Matrix multiplication is what converts the derivatives of f and g into that of their composition. This is not a surprising result – we saw in Week 4 that composition of linear transformations corresponds to multiplication of the matrices together (in the correct order!). What is surprising is that 20th century calculus texts did not use matrices to explain the Chain Rule, relying instead on memorization of many different formulae depending on the number of inputs and outputs.

The Chain Rule can be used to derive most other interesting differentiation rules, such as the product rules for dot products and cross products from Week 2. In fact, the Chain Rule can use simple matrix multiplication to re-derive the classical product rule for single variable functions, should one want to do such a thing. More advanced applications (such as the *material derivative* from Mechanics) are possible. Though not essential to our story, this is a quick sidequest. Consider an elastic or fluid substance of particles (a continuum of atoms, if one prefers) at positions $\boldsymbol{x}(t)$ (usually in $\mathbb{R}^2$ or $\mathbb{R}^3$) where t is time – the substance is bending or flowing. A time-dependent function $h(\boldsymbol{x}, t)$, perhaps representing pressure or temperature of the substance, is in fact a composition, since $\boldsymbol{x} = \boldsymbol{x}(t)$. The *material derivative* of h measures how h changes in time, from the perspective of a moving particle. In Physics, this has a special notation:

$$\frac{D}{Dt}h = \frac{\partial h}{\partial t} + \frac{\partial h}{\partial \boldsymbol{x}}\frac{\partial \boldsymbol{x}}{\partial t}.$$

This is, of course, just the Chain Rule in action for h as a function of t. (See the Epilogue for the use of this notation in fluid dynamics.)

THE INVERSE FUNCTION THEOREM. Recall from pre-calculus the notion of the inverse of a function of one variable. One says $f: \mathbb{R} \to \mathbb{R}$ is invertible if there exists an inverse, denoted f^{-1}, such that $f^{-1}(f(x)) = x$ and $f(f^{-1}(y)) = y$ for all x and y. The same definition holds in the multivariate case where f and f^{-1} have domain and range in $\mathbb{R}^n$ (the number of variables *must* be equal). The computation of an inverse (much less its existence!) is difficult in all but the simplest settings. For a linear function, $f(\boldsymbol{x}) = A\boldsymbol{x}$, where A is an n-by-n matrix, we know that the inverse function $f^{-1}(\boldsymbol{y}) = A^{-1}\boldsymbol{y}$ exists only when $\det A \neq 0$ (thus guaranteeing the inverse matrix A^{-1} exists). This points the way to a deep theorem about the existence of an inverse function.

The *Inverse Function Theorem* states that a continuously differentiable function $f: \mathbb{R}^n \to \mathbb{R}^n$ is guaranteed to be *locally invertible* at input $\boldsymbol{a}$ and output $f(\boldsymbol{a})$ if the derivative of f at a is invertible; that is, if $\det[Df]_a \neq 0$. Furthermore, if the inverse exists, its derivative is $[Df^{-1}]_{f(a)} = [Df]_a^{-1}$. This follows from the Chain Rule applied to the definition of an inverse:

$$f \circ f^{-1} = id = f^{-1} \circ f \quad \Rightarrow \quad [Df][Df^{-1}] = I = [Df^{-1}][Df].$$

There are a few subtleties here. The first is the sufficiency of the criterion. If the derivative is invertible, then the function is [locally] invertible as well; if the derivative is not invertible, then the criterion fails, and more work is needed to determine invertibility. It is not an if-and-only-if condition. This can be seen from the simple single-variable example of $f(x) = x^3$, whose inverse $f^{-1}(y) = \sqrt[3]{y}$ exists, even though $f'(0) = 0$.

The second and more difficult subtlety is the local nature of the theorem. The function $f(x) = x^2$ is locally invertible about any point that is nonzero, though the size of the domain on which invertibility holds shrinks as one approaches a zero derivative. It is fantastic that the Inverse Function Theorem can guarantee invertibility of the fully nonlinear function based only on linear data: the price of that power is a loss of certainty about the region on which invertibility holds.

THE IMPLICIT FUNCTION THEOREM. The Inverse Function Theorem is the shadow of a much deeper and more fundamental result. Consider an implicit equation (or, really, m equations) of the form

$$F(\boldsymbol{x},\boldsymbol{y}) = \boldsymbol{0}\,,$$

where $F\colon \mathbb{R}^{n+m} \to \mathbb{R}^m$ with input variables split into $\boldsymbol{x} \in \mathbb{R}^n$ and $\boldsymbol{y} \in \mathbb{R}^m$. We say that one can solve implicitly for the $\boldsymbol{y}$ variables in terms of the $\boldsymbol{x}$ variables if there is a function $\boldsymbol{y} = \boldsymbol{y}(\boldsymbol{x})$ which satisfies the equation $F(\boldsymbol{x},\boldsymbol{y}(\boldsymbol{x})) = \boldsymbol{0}$. When is this possible? Not always, as one recalls from such simple equations as $x^2 + y^2 - 1 = 0$. The Implicit Function Theorem says the one can solve for $\boldsymbol{y} = \boldsymbol{y}(\boldsymbol{x})$ locally, about some input $\boldsymbol{a}$, if a certain square submatrix of the derivative $[DF]$ is invertible, *i.e.*:

$$\det\left[\frac{\partial F}{\partial \boldsymbol{y}}\right] \neq 0\,.$$

In this case, one is guaranteed to have a *local* solution $\boldsymbol{y} = \boldsymbol{y}(\boldsymbol{x})$ whose derivative is:

$$\left[\frac{\partial \boldsymbol{y}}{\partial \boldsymbol{x}}\right] = -\left[\frac{\partial F}{\partial \boldsymbol{y}}\right]^{-1}\left[\frac{\partial F}{\partial \boldsymbol{x}}\right].$$

This is extremely useful, as it allows one to linearly approximate a solution to a set of nonlinear equations with simple partial derivative data and matrix algebra. Applications of this result range from Economics to GPS and much more: we will use the IFT in two weeks when doing constrained optimization.

DISCUSSION

QUESTION 1. Consider the following three derivatives of functions, each of which takes the origin to the origin:

$$[Df]_{\boldsymbol{0}} = \begin{bmatrix} 1 & -3 & 0 \\ 2 & 7 & 1 \\ 3 & 1 & 0 \\ -2 & 1 & -1 \end{bmatrix} \; : \; [Dg]_{\boldsymbol{0}} = \begin{bmatrix} 4 & 1 \\ -2 & 0 \\ -1 & 1 \end{bmatrix} \; : \; [Dh]_{\boldsymbol{0}} = \begin{bmatrix} 3 & -1 & 2 & 0 \\ -2 & -1 & 1 & 4 \\ 5 & 0 & 4 & 2 \end{bmatrix}$$

Which derivatives (at the origin) can you compute among:

$$f\circ g \;:\; g\circ f \;:\; g\circ h \;:\; h\circ g \;:\; f\circ h \;:\; h\circ f$$

This is a good time to emphasize that you can tell a lot about a function by knowing its derivative at a single point. How many inputs and how many outputs do these functions have? Why did we have to specify that these functions all take the origin to the origin? Wait, is this the same origin?

QUESTION 2. Do you remember your differentiation rules? Let's recall... [*have students recall the product and quotient rules...*] Are there any other rules you remember? [*students will likely recall the inverse rule, so go ahead and see how that follows from the Chain Rule...*] What else follows from the Chain Rule? What about:

$$\left(u(x)^{v(x)}\right)' = u'vu^{v-1} + v'u^v \ln u$$

Use a simple matrix product to derive this result, using $f(u,v) = u^v$ and $g(x) = (u(x), v(x))^T$. Having just reviewed some partial derivatives, the rest is straightforward. It is worth reminding students that memorizing this formula is not productive – once you internalize the Chain Rule.

QUESTION 3. What is the derivative of the square $f(\boldsymbol{x}) = Q^2$ of the quadratic form $Q(\boldsymbol{x}) = \boldsymbol{x} \cdot A\boldsymbol{x}$?

The derivative of Q was computed in the videolectures to be $[DQ] = \boldsymbol{x}^T(A + A^T)$. This problem makes students very uncomfortable – the notation is unfamiliar. If this happens, encourage them to work out a simple example and verify what the notation means.

QUESTION 4. You might remember from single-variable calculus using the Chain Rule together with the Fundamental Theorem of Integral Calculus to compute derivatives of integrals with respect to a variable appearing in the limits of integration. Use the multivariate Chain Rule to redo this, computing $[DF]$ where:

$$F(x) = \int_{g(x)}^{h(x)} f(t)\, dt$$

Is this a multivariable problem at all? Does matrix multiplication help here? This is a good problem for generating discussion, as well as recalling the importance of integrals and the FTIC.

QUESTION 5. Consider the function $F: \mathbb{R}^3 \to \mathbb{R}^3$ given by

$$F\begin{pmatrix} x \\ y \\ z \end{pmatrix} = \begin{pmatrix} u \\ v \\ w \end{pmatrix} = \begin{pmatrix} \arctan(x+y) \\ 3x - 2z \\ 1 + e^{y+z} \end{pmatrix}.$$

Is this function locally invertible near $x = y = z = 0$? Is it everywhere locally invertible? Is it in fact invertible?

Students often struggle with the logical details of a sufficient-but-not-necessary criterion. Thinking in terms of domain and range (or codomain) is helpful in this particular case. For a twist, replace the arctangent function with a cube root.

QUESTION 6. The following are three implicitly defined surfaces:

$$x^2 + y^2 + z^2 = C_1 \quad : \quad z - \cosh x - \cosh y = C_2 \quad : \quad x - 2yz + y^3 = C_3$$

These surfaces all intersect at the point $(1,1,2)$ when $C_1 = \sqrt{6}$, $C_2 = 2 - e - 1/e$, and $C_3 = -2$. What does this point of mutual intersection do when you wiggle the three constants $\{C_i\}$? Does it fill in a neighborhood of $(1,1,2)$? Or does it trace out some lower-dimensional set?

This certainly seems like a strange question that has nothing to do with this week's material. Try to get students to think of the C_i as variables $\boldsymbol{C}$ and the point of common intersection $\boldsymbol{x} = (x,y,z)^T$ in terms of a function $\boldsymbol{C} = F(\boldsymbol{x})$. What does the Inverse Function Theorem mean in this setting? What does local invertibility imply about the intersection point? Where should the derivative $[DF]$ be evaluated?

QUESTION 7. Let's practice the Implicit Function Theorem, starting with the 1-D case. Where can we solve for $y = y(x)$ given that $xe^y - ye^x = 1$?

For the Implicit Function Theorem, students will be intimidated: review the statement that for $F(\boldsymbol{x},\boldsymbol{y}) = \boldsymbol{0}$, we have $\left[\frac{\partial y}{\partial x}\right] = \left[\frac{\partial F}{\partial y}\right]^{-1}\left[\frac{\partial F}{\partial x}\right]$ if the inverse exists, beginning with the 1-D version.

QUESTION 8. A fully nonlinear problem: can you solve for c, d as a function of a, b, given that $ab - bc + cd = 8$ and $a + 2b - 2c + 4d = 12$, assuming that you are near $a = 1, b = 2, c = 3, d = 4$.

What does this problem mean geometrically? What does the solution to these equations look like in 4-D? Try to get students to think in terms of dimension and degrees of freedom (if only we had a name for this quantity... that's for next semester!)

QUESTION 9. Now, repeat the last problem in the linear case. Assuming that you are near $a = 0, b = 1, c = -2, d = 2$ solve for c, d if

$$a + 2b - 3c + 4d = 16$$
$$-2a + b + c - 3d = -7$$

Do this via the IFT, then do it explicitly using row reduction. Ahha! Row reduction and back-substitution is really the linear version of what the IFT is doing...

QUESTION 10. Recall the infinite power tower, defined implicitly via $y = x^y$. In single-variable calculus, one uses logarithmic differentiation to show that the derivative (where it exists) equals

$$\frac{dy}{dx} = \frac{y^2}{x(1 - \ln x)}$$

The derivative does not exist everywhere, because we cannot solve for $y = y(x)$ everywhere given the implicit equation $y = x^y$. What does the Implicit Function Theorem say about this situation?

This should cause some perhaps frustrated or confused discussion. What is the function $F(x,y) = 0$ to be used? Try to get students to remember that the IFT is a local result and requires evaluation at a particular point. In this case, constraining $\partial F/\partial y$ is doable: $1 - x^y \ln x \neq 0$. But what does this mean in terms of bounds on x? Clearly $x = y = 1$ is a solution. Methods outside the bounds of this course can give a precise interval of convergence: $e^{-e} \leq x \leq e^{1/e}$.

QUESTION 11. Consider the following multivariate infinite power tower defined by $z(x,y) = x^{y^z}$. This is like the original infinite power tower, but with two inputs (x and y) with alternating powers. Are there any values of x and y at which this function can be said to exist, and can you compute its derivative?

This is an unreasonably intimidating and difficult problem: not recommended unless students want a real challenge. Still, it is doable with enough labor.

QUESTION 12. Can you show that the Inverse Function Theorem is a special case of the Implicit function theorem?

Try starting with a recollection of inverse and implicit functions. Because of the confusion possible between variable names, it might be best to start with $\mathbf{u} = f(\mathbf{v})$ and look for a putative inverse of the form $\mathbf{v} = g(\mathbf{u})$. Beginning with the equations $F(\mathbf{u}, \mathbf{v}) = \mathbf{u} - f(\mathbf{v}) = \mathbf{0}$, then, what would it mean to have $\mathbf{v} = g(\mathbf{u})$ satisfying $\mathbf{u} - \mathbf{f}(\mathbf{g}(\mathbf{u})) = \mathbf{0}$? In this case, what derivatives do you have to check? Remember, an inverse must also satisfy $\mathbf{g}(\mathbf{f}(\mathbf{v})) = \mathbf{v}$: does this cause any trouble?

ASSESSMENT PROBLEMS

PROBLEM 1. Consider the functions

$$f\begin{pmatrix} x \\ y \\ z \end{pmatrix} = \begin{pmatrix} x^2 + yz^2 \\ 2x + y^3 - z \end{pmatrix} \quad : \quad g\begin{pmatrix} u \\ v \end{pmatrix} = \begin{pmatrix} u^2 - v^2 \\ uv \\ 3u - 2v \end{pmatrix}$$

A) Compute the derivatives of f and g.

B) Using the Chain Rule, compute the derivative of the composition $f \circ g$ at the point where all its inputs equal $+1$.

C) If all inputs of $f \circ g$ equal $+1$ and are decreasing at a unit rate, at what rate is the first output of $f \circ g$ changing?

PROBLEM 2. Consider the following functions:

$$f\begin{pmatrix} a \\ b \\ c \end{pmatrix} = \begin{pmatrix} \ln(ab) \\ abc \\ b^2 + 5c \end{pmatrix} \quad : \quad g\begin{pmatrix} x \\ y \\ z \end{pmatrix} = \begin{pmatrix} x^2 + y^2 z \\ 2x - y + 3z \\ 3x - 4y \end{pmatrix}$$

A) Compute the derivative $[Dg]$ and evaluate at $x = 1, y = 0, z = 0$.

B) Compute the derivative $[Df]$.

C) Use the Chain Rule to compute $[D(f \circ g)]$ evaluated at $x = 1, y = 0, z = 0$.

PROBLEM 3. Consider the following functions:

$$f\begin{pmatrix} u \\ v \end{pmatrix} = \begin{pmatrix} 2u - 3v \\ (u - v)^{-1} \\ uv \end{pmatrix} \quad : \quad g\begin{pmatrix} x \\ y \end{pmatrix} = \begin{pmatrix} x + 2y \\ (x - 3y)^2 \end{pmatrix}$$

A) Compute the derivative $[Df]$.

B) Compute the derivative $[Dg]$.

C) Use the Chain Rule to compute the derivative of $f \circ g$ evaluated at $(1,1)$.

PROBLEM 4. Consider the following functions:

$$h\begin{pmatrix} s \\ t \end{pmatrix} = \begin{pmatrix} t^2 - 4s \\ \ln(1 - s) \\ e^{s - t^2} \end{pmatrix} \quad : \quad g\begin{pmatrix} x \\ y \end{pmatrix} = \begin{pmatrix} 3x + 2y \\ 1 - x + y^2 \end{pmatrix}$$

A) Compute the derivative $[Dh]$.

B) Compute the derivative $[Dg]$.

C) Use the Chain Rule to compute the derivative of $h \circ g$ at the origin.

PROBLEM 5. There are three functions, $f, g, \& h$, each of which sends the origin to the origin. At the origin, these functions have derivatives equal to:

$$[Df]_0 = \begin{bmatrix} 1 & 2 & 0 & 4 \\ 2 & 0 & -1 & 0 \\ 0 & 3 & -3 & -2 \end{bmatrix} \quad : \quad [Dg]_0 = \begin{bmatrix} -1 & 0 & 2 \\ 0 & 2 & 0 \\ -3 & 5 & 1 \\ 2 & 0 & 1 \end{bmatrix} \quad : \quad [Dh]_0 = \begin{bmatrix} 1 & 0 \\ 3 & 2 \\ -2 & -1 \end{bmatrix}$$

A) How many inputs and outputs does the function $f \circ g \circ h = f(g(h))$ have?

B) Compute the derivative at the origin, $[D(f \circ g \circ h)]_0$.

C) If, at the origin, the inputs of $f \circ g \circ h$ are all changing at the same rate, which output is most sensitive to the change?

PROBLEM 6. There are three differentiable functions, $f, g, \& h$, each of which sends the origin to the origin. At the origin, these functions have derivatives equal to:

$$[Df]_0 = \begin{bmatrix} 2 & 0 & 2 \\ 0 & 1 & 0 \\ 3 & 1 & 0 \\ 0 & 0 & 4 \end{bmatrix} \quad : \quad [Dg]_0 = \begin{bmatrix} -1 & 0 & 3 \\ 0 & 1 & -6 \\ 7 & 5 & -2 \\ 2 & 0 & 1 \end{bmatrix} \quad : \quad [Dh]_0 = \begin{bmatrix} 1 & 2 & 0 & 1 \\ 2 & 0 & -1 & 0 \\ 0 & 3 & 0 & 2 \end{bmatrix}$$

A) Which of the following compositions are not legal?

$$(f \circ g) \quad : \quad (g \circ g) \quad : \quad (g \circ h) \quad : \quad (h \circ g) \quad : \quad (f \circ h) \quad : \quad (h \circ h)$$

B) Compute the derivative at the origin, $[D(g-3f)]_0$.

C) Compute the derivative of $h \circ f$ at the origin.

PROBLEM 7. There are three differentiable functions, $f, g, \& h$, each of which sends the origin to the origin. At the origin, these functions have derivatives equal to:

$$[Df]_0 = \begin{bmatrix} 1 & 2 & 0 & 4 \\ 2 & 0 & -1 & 0 \\ 0 & 3 & -3 & -2 \end{bmatrix} \quad : \quad [Dg]_0 = \begin{bmatrix} -1 & 0 & 2 \\ 0 & 2 & 0 \\ -3 & 5 & 1 \\ 2 & 0 & 1 \end{bmatrix} \quad : \quad [Dh]_0 = \begin{bmatrix} 7 & -9 \\ 5 & 3 \\ 2 & -4 \end{bmatrix}$$

A) Which compositions are not legal?

$$(f \circ g) \ : \ (g \circ f) \ : \ (g \circ h) \ : \ (h \circ g) \ : \ (f \circ h) \ : \ (h \circ f)$$

B) Compute the derivative at the origin, $[D(g \circ f)]_0$.

C) If, at the origin, the first input of $g \circ f$ is increasing and the remaining inputs are decreasing at the same rate, which output is most sensitive to the change?

PROBLEM 8. Consider the following functions:

$$f\begin{pmatrix} u \\ v \\ w \end{pmatrix} = \begin{pmatrix} u^2 - 3vw \\ uv^2 - uvw \end{pmatrix} = \begin{pmatrix} s \\ t \end{pmatrix} \quad : \quad g\begin{pmatrix} x \\ y \end{pmatrix} = \begin{pmatrix} x - y \\ x^2 - y^2 \\ 2x - 3y \end{pmatrix} = \begin{pmatrix} a \\ b \\ c \end{pmatrix}$$

A) Compute the derivatives $[Df]$ and $[Dg]$.

B) What are the input and output variables of the composition $f \circ g$?

C) Compute the derivative of the composition $D[f \circ g]$ at the point where all inputs are equal to 2.

D) Is the composition $f \circ g$ invertible locally where all the inputs equal 2?

PROBLEM 9. There are three functions, $f, g, \& h$, each of which sends the origin to the origin. At the origin, these functions have derivatives equal to:

$$[Df]_0 = \begin{bmatrix} 3 & 0 & 1 & 0 \\ 0 & -1 & 2 & -1 \end{bmatrix} \quad : \quad [Dg]_0 = \begin{bmatrix} 0 & 2 \\ 4 & 3 \\ 1 & -1 \\ 0 & -3 \end{bmatrix} \quad : \quad [Dh]_0 = \begin{bmatrix} 1 & 3 \\ -2 & 0 \end{bmatrix}$$

A) How many inputs and outputs does the function $f \circ g \circ h = f(g(h))$ have?

B) Compute the derivative at the origin of the composition $f \circ g \circ h$.

C) Compute the derivative at the origin of h^{-1}, the inverse of h.

PROBLEM 10. Consider the functions

$$f\begin{pmatrix} u \\ v \\ w \end{pmatrix} = \begin{pmatrix} u^2 - 3v \\ uv - w^2 \end{pmatrix} \quad : \quad g\begin{pmatrix} x \\ y \end{pmatrix} = \begin{pmatrix} xy^2 \\ x^2 - y \\ 3y \end{pmatrix}$$

A) Compute the derivatives of f and g.

B) Compute the derivative of the composition $f \circ g$ at the point where all its inputs equal $+1$.

C) If all inputs of $f \circ g$ equal $+1$ and are decreasing at a unit rate, at what rate is the last output changing?

PROBLEM 11. Consider the following functions:

$$f\begin{pmatrix}x\\y\\z\end{pmatrix}=\begin{pmatrix}x^2y-2z\\x-y^3+z^2\\x-y+z\end{pmatrix}\quad:\quad g\begin{pmatrix}u\\v\\w\end{pmatrix}=\begin{pmatrix}2u\\v+w\\2v+3w\end{pmatrix}$$

A) Compute the derivatives $[Df]$ and $[Dg]$ evaluated at the origin.

B) Compute the derivative of f composed with g at the origin, $[D(f\circ g)]_{\mathbf{0}}$.

C) Explain: how would you compute the derivative of f^{-1}, the inverse of f, at the origin?

PROBLEM 12. Is the function

$$G\begin{pmatrix}x\\y\\u\\v\end{pmatrix}=\begin{pmatrix}2u+3v\\v+(1-u^2)^{-1}\\2y-\sin x\\e^x-\cos 3x+\ln(1-2y)\end{pmatrix}$$

locally invertible near the origin? Explain.

PROBLEM 13. Is the function

$$F\begin{pmatrix}x\\y\\z\\t\end{pmatrix}=\begin{pmatrix}2t-\sin z\\z-e^t-1\\e^{3x}-2y-1\\2x+\ln(1+y)\end{pmatrix}$$

locally invertible near the origin?

PROBLEM 14. Consider the functions

$$f\begin{pmatrix}x\\y\\z\end{pmatrix}=\begin{pmatrix}u\\v\end{pmatrix}=\begin{pmatrix}xy+yz\\xz+yz\end{pmatrix}\quad:\quad g\begin{pmatrix}s\\t\end{pmatrix}=\begin{pmatrix}x\\y\\z\end{pmatrix}=\begin{pmatrix}s^2\cos t\\s^2\sin t\\s\end{pmatrix}$$

A) Compute the derivatives of f and g

B) Argue carefully that one can solve for (s,t) as functions of (u,v) when (s,t) is close to the point $\left(2,\frac{\pi}{2}\right)$ and (u,v) is close to $(8,8)$.

PROBLEM 15. Consider the function

$$f\begin{pmatrix}x\\y\\z\end{pmatrix}=\begin{pmatrix}u\\v\\w\end{pmatrix}=\begin{pmatrix}e^{x+2y}-1\\1+3x-\cos 2y\\\sin(3x+y^3-2z)\end{pmatrix}$$

A) Compute the derivative $[Df]$ evaluated at the origin.

B) Explain: what does the Inverse Function Theorem say about the invertibility of this f near the origin?

PROBLEM 16. Consider the equations

$$\begin{aligned}2uv&=\sin(x-2y)\\4x+y^2&=3u+e^v\end{aligned}$$

A) Write these equations in the form $F=0$ for F some function of 4 inputs and 2 outputs. What is F?

B) Use the Implicit Function Theorem to show that near the origin, one can solve for $x=x(u,v)$ and $y=y(u,v)$ while satisfying these equations.

ANSWERS & HINTS

PROBLEM 1. A) $[Df] = \begin{bmatrix} 2x & z^2 & 2yz \\ 2 & 3y^2 & -1 \end{bmatrix}, [Dg] = \begin{bmatrix} 2u & -2v \\ v & u \\ 3 & -2 \end{bmatrix}$; B) $g\begin{pmatrix}1\\1\end{pmatrix} = \begin{pmatrix}0\\1\\1\end{pmatrix}$ so that $[D(f \circ g)]_{1,1} = \begin{bmatrix} 0 & 1 & 2 \\ 2 & 3 & -1 \end{bmatrix}\begin{bmatrix} 2 & -2 \\ 1 & 1 \\ 3 & -2 \end{bmatrix} = \begin{bmatrix} 7 & -3 \\ 4 & 1 \end{bmatrix}$; C) $-7-(-3) = -4$

PROBLEM 2. A) $[Dg]_0 = \begin{bmatrix} 2 & 0 & 0 \\ 2 & -1 & 3 \\ 3 & -4 & 0 \end{bmatrix}$; B) $[Df] = \begin{bmatrix} b/a & a/b & 0 \\ bc & ab & ab \\ 0 & 2b & 5 \end{bmatrix}$; C) $g\begin{pmatrix}1\\0\\0\end{pmatrix} = \begin{pmatrix}1\\2\\3\end{pmatrix}$, so by the Chain Rule, $[D(f \circ g)]_{1,0,0} = \begin{bmatrix} 1 & 1/2 & 0 \\ 6 & 3 & 2 \\ 0 & 4 & 5 \end{bmatrix}\begin{bmatrix} 2 & 0 & 0 \\ 2 & -1 & 3 \\ 3 & -4 & 0 \end{bmatrix} = \begin{bmatrix} 3 & -1/2 & 3/2 \\ 24 & -11 & 9 \\ 23 & -24 & 12 \end{bmatrix}$

PROBLEM 3. A) $[Df] = \begin{bmatrix} 2 & -3 \\ -(u-v)^{-2} & (u-v)^{-2} \\ v & u \end{bmatrix}$; B) $[Dg] = \begin{bmatrix} 1 & 2 \\ 2(x-3y) & -6(x-3y) \end{bmatrix}$; C) since $g\begin{pmatrix}1\\1\end{pmatrix} = \begin{pmatrix}3\\4\end{pmatrix}$, by the Chain Rule,

$$[D(f \circ g)]_{1,1} = \begin{bmatrix} 2 & -3 \\ -(-1)^{-2} & (-1)^{-2} \\ 4 & 3 \end{bmatrix}\begin{bmatrix} 1 & 2 \\ -4 & 12 \end{bmatrix} = \begin{bmatrix} 14 & -32 \\ -5 & 10 \\ -8 & 44 \end{bmatrix}$$

PROBLEM 4. A) $[Dh] = \begin{bmatrix} -4 & 2t \\ -(1-s)^{-1} & 0 \\ e^{s-t^2} & -2te^{s-t^2} \end{bmatrix}$; B) $[Dg] = \begin{bmatrix} 3 & 2 \\ -1 & 2y \end{bmatrix}$; C) $g\begin{pmatrix}0\\0\end{pmatrix} = \begin{pmatrix}0\\1\end{pmatrix}$, so, by the Chain Rule $[D(h \circ g)]_0 = \begin{bmatrix} -4 & 2 \\ -1 & 0 \\ 1/e & -2/e \end{bmatrix}\begin{bmatrix} 3 & 2 \\ -1 & 2y \end{bmatrix} = \begin{bmatrix} -14 & -8 \\ -3 & -2 \\ 5/e & 2/e \end{bmatrix}$

PROBLEM 5. A) 2 inputs, 3 outputs ; B) $\begin{bmatrix} 7 & 2 \\ -20 & -13 \\ -12 & -13 \end{bmatrix}$; C) 2nd output

PROBLEM 6. A) $f \circ g, g \circ g, h \circ h$ illegal ; B) $\begin{bmatrix} -7 & 0 & -3 \\ 0 & -2 & -6 \\ -2 & 2 & -2 \\ 2 & 0 & -11 \end{bmatrix}$; C) $\begin{bmatrix} 2 & 2 & 6 \\ 1 & -1 & 4 \\ 0 & 3 & 8 \end{bmatrix}$

PROBLEM 7. A) $h \circ g, f \circ h, h \circ f$ illegal ; B) $\begin{bmatrix} -1 & 4 & -6 & -8 \\ 4 & 0 & -2 & 0 \\ 7 & -3 & -8 & -14 \\ 2 & 7 & -3 & 6 \end{bmatrix}$; C) 3rd output

PROBLEM 8. A) $[Df] = \begin{bmatrix} 2u & -3w & -3v \\ v^2 - vw & 2uv - uw & -uv \end{bmatrix}, [Dg] = \begin{bmatrix} 1 & -1 \\ 2x & -2y \\ 2 & -3 \end{bmatrix}$; B) $\begin{pmatrix}x\\y\end{pmatrix} \mapsto \begin{pmatrix}s\\t\end{pmatrix}$; C) $g\begin{pmatrix}2\\2\end{pmatrix} = \begin{pmatrix}0\\0\\-2\end{pmatrix}$, so $[D(f \circ g)]_{2,2} = \begin{bmatrix} 0 & 6 & 0 \\ 0 & 0 & 0 \end{bmatrix}\begin{bmatrix} 1 & -1 \\ 4 & -4 \\ 2 & -3 \end{bmatrix} = \begin{bmatrix} 24 & -24 \\ 0 & 0 \end{bmatrix}$; D) maybe, maybe not – the Inverse Function Theorem is inconclusive

PROBLEM 9. A) 2 inputs, 2 outputs ; B) $\begin{bmatrix} -9 & 3 \\ 2 & -6 \end{bmatrix}$; C) $[Dh^{-1}]_0 = [Dh]_0^{-1} = \frac{1}{6}\begin{bmatrix} 0 & -3 \\ 2 & 1 \end{bmatrix}$ (note that h sends the origin to itself, which is important)

PROBLEM 10. A) $[Df] = \begin{bmatrix} 2u & -3 & 0 \\ v & u & -2w \end{bmatrix}$ and $[Dg] = \begin{bmatrix} y^2 & x \\ 2x & -1 \\ 0 & 3 \end{bmatrix}$; B) $[Dg]_{1,1} = \begin{bmatrix} 1 & 1 \\ 2 & -1 \\ 0 & 3 \end{bmatrix}$ and $g\begin{pmatrix}1\\1\end{pmatrix} = \begin{pmatrix}1\\0\\3\end{pmatrix}$, so $[Df]_{1,0,3} = \begin{bmatrix} 2 & -3 & 0 \\ 0 & 1 & -6 \end{bmatrix}$ and, via the Chain Rule, $[D(f \circ g)]_{1,1} = \begin{bmatrix} 2 & -3 & 0 \\ 0 & 1 & -6 \end{bmatrix}\begin{bmatrix} 1 & 1 \\ 2 & -1 \\ 0 & 3 \end{bmatrix} = \begin{bmatrix} -4 & 5 \\ 2 & -19 \end{bmatrix}$; C) $[D(f \circ g)]_{1,1}\begin{pmatrix}-1\\-1\end{pmatrix} = \begin{bmatrix} -4 & 5 \\ 2 & -19 \end{bmatrix}\begin{pmatrix}-1\\-1\end{pmatrix} = \begin{pmatrix}-1\\17\end{pmatrix}$

PROBLEM 11. A) $[Df]_0 = \begin{bmatrix} 0 & 0 & -2 \\ 1 & 0 & 0 \\ 1 & -1 & 1 \end{bmatrix}$ and $[Dg]_0 = \begin{bmatrix} 2 & 0 & 0 \\ 0 & 1 & 1 \\ 0 & 2 & 3 \end{bmatrix}$; B) by the Chain Rule, $[D(f \circ g)]_0 = \begin{bmatrix} 0 & 0 & -2 \\ 1 & 0 & 0 \\ 1 & -1 & 1 \end{bmatrix}\begin{bmatrix} 2 & 0 & 0 \\ 0 & 1 & 1 \\ 0 & 2 & 3 \end{bmatrix} = \begin{bmatrix} 0 & -4 & -6 \\ 2 & 0 & 0 \\ 2 & 1 & 2 \end{bmatrix}$; C) $[Df^{-1}]_0 = [Df]_0^{-1}$ (note that f sends the origin to itself, which is important)

PROBLEM 12. $[DG]_0 = \begin{bmatrix} 0 & 0 & 2 & 3 \\ 0 & 0 & 0 & 1 \\ -1 & 2 & 0 & 0 \\ 1 & -2 & 0 & 0 \end{bmatrix}$ which has determinant $\begin{vmatrix} 2 & 3 \\ 0 & 1 \end{vmatrix}\begin{vmatrix} -1 & 2 \\ 1 & -2 \end{vmatrix} = 0.$ Inverse Function Theorem fails and invertibility is uncertain

PROBLEM 13. $[DF]_0 = \begin{bmatrix} 0 & 0 & -1 & 2 \\ 0 & 0 & 1 & -1 \\ 3 & -2 & 0 & 0 \\ 2 & 1 & 0 & 0 \end{bmatrix}$ which has determinant $\begin{vmatrix} 3 & -2 \\ 2 & 1 \end{vmatrix}\begin{vmatrix} -1 & 2 \\ 1 & -1 \end{vmatrix} = -7 \neq 0$. Inverse Function Theorem implies local invertibility near the origin

PROBLEM 14. A) $[Df] = \begin{bmatrix} y & x+z & y \\ z & z & x+y \end{bmatrix}, [Dg] = \begin{bmatrix} 2s\cos t & -s^2 \sin t \\ 2s \sin t & s^2 \cos t \\ 1 & 0 \end{bmatrix}$; B) since $g\begin{pmatrix} 2 \\ \frac{\pi}{2} \end{pmatrix} = \begin{pmatrix} 0 \\ 4 \\ 2 \end{pmatrix}$, evaluate $[Df]_{0,4,2} = \begin{bmatrix} 4 & 2 & 4 \\ 2 & 2 & 4 \end{bmatrix}$ and $[Dg]_{2,\frac{\pi}{2}} = \begin{bmatrix} 0 & -4 \\ 4 & 0 \\ 1 & 0 \end{bmatrix}$ using the Chain Rule to determine $[D(f \circ g)]_{2,\frac{\pi}{2}} = \begin{bmatrix} 4 & 2 & 4 \\ 2 & 2 & 4 \end{bmatrix}\begin{bmatrix} 0 & -4 \\ 4 & 0 \\ 1 & 0 \end{bmatrix} = \begin{bmatrix} 12 & -16 \\ 12 & -8 \end{bmatrix}$, which has determinant non-zero; invertibility follows from the Inverse Function Theorem

PROBLEM 15. A) $[Df]_0 = \begin{bmatrix} 1 & 2 & 0 \\ 3 & 0 & 0 \\ 3 & 0 & -2 \end{bmatrix}$; B) $\det[Df]_0 = 6$, thus, locally invertible

PROBLEM 16. The function is $F = \begin{pmatrix} 2uv - \sin(x - 2y) \\ 4x + y^2 - 3u - e^v \end{pmatrix}$ and its derivative with respect to the (x, y) variables at the origin is $\begin{bmatrix} -1 & 2 \\ 4 & 0 \end{bmatrix}$, the determinant of which is nonzero.

Week 7 : Approximation

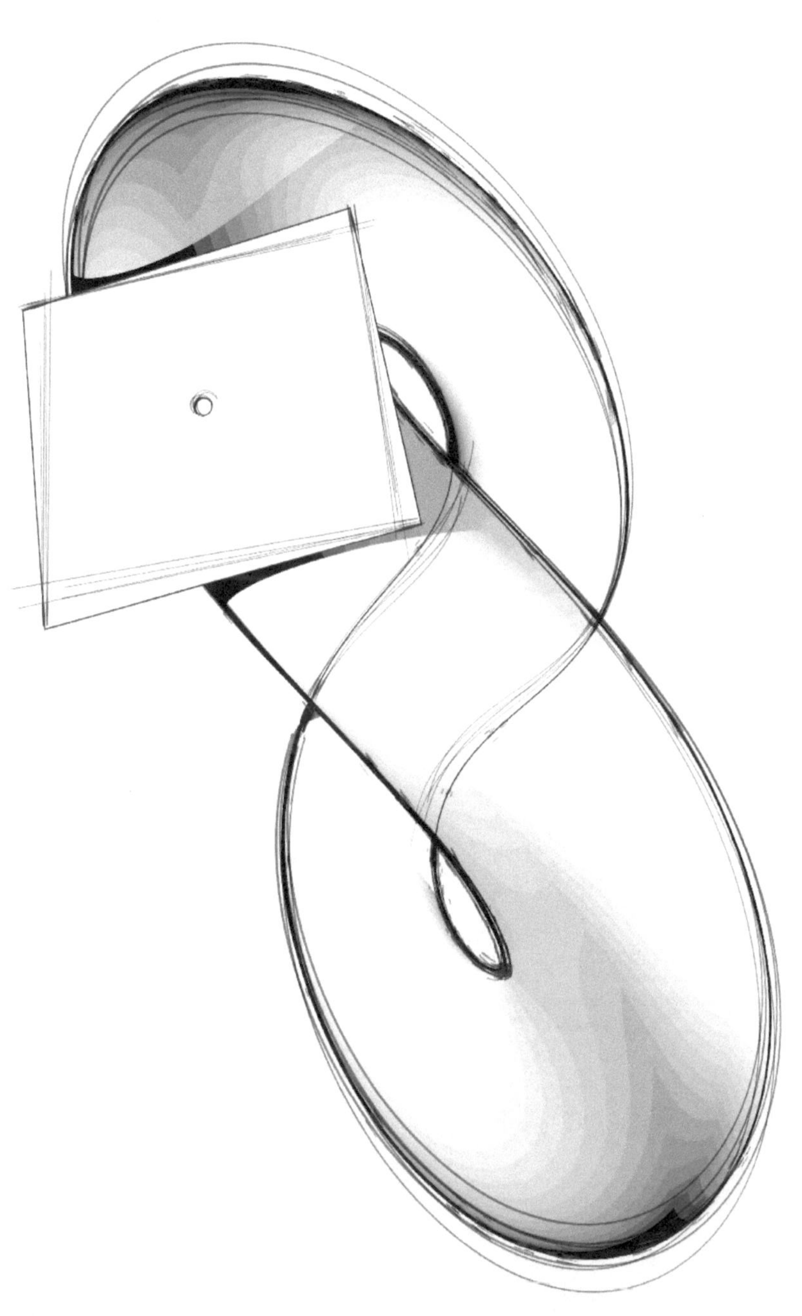

OUTLINE

MATERIALS: Calculus BLUE : Vol 2 : Chapters 9-13

TOPICS:

- ▷ Level sets of scalar-valued functions
- ▷ Gradients of scalar-valued functions
- ▷ Tangent planes to surfaces via the derivative/gradient
- ▷ Differentials and approximations
- ▷ Relative rates of change and approximations via linearization
- ▷ Taylor expansion as polynomial approximation
- ▷ Multi-index notation for Taylor expansion
- ▷ Mixed higher-order partial derivatives
- ▷ The second derivative [Hessian] of a scalar-valued function as a matrix

LEARNING OBJECTIVES:

- ▷ Use the notation for level sets and describe/draw simple level sets
- ▷ Compute gradients of scalar-valued functions
- ▷ Relate gradients, derivatives, and differentials
- ▷ Compute tangent planes to implicit and parametrized surfaces
- ▷ Compute df for a scalar-valued f via implicit differentiation
- ▷ Compute and interpret relative rates of change via differentials
- ▷ Linearly approximate multivariate functions via differentials
- ▷ Recognize and use multi-index notation in the context of Taylor series
- ▷ Compose single-variable Taylor series to expand multivariate functions
- ▷ Organize terms in multivariate Taylor series by degree
- ▷ Determine partial derivatives of a function based on its Taylor expansion

PRIMER

This is the week where we return to Geometry as a source of meaning and intuition for derivatives.

LEVEL SETS. For the next several weeks, we will restrict attention to functions that have a single output. We will sometimes (though more frequently in Weeks 12-14) call these *scalar fields*: a scalar is assigned to every point. Such scalar-valued functions can be more easily visualized than their multi-valued counterparts by thinking in terms of *level sets*.

The level sets of a scalar-valued function $f: \mathbb{R}^n \to \mathbb{R}$ are subsets of the domain on which f is a fixed value. One can think of constant-temperature curves (*isotherms*) on a weather map or perhaps constant-height (*contour*) curves on a topographic map. The notation for a level set can seem unusual, as it resembles an inverse (which of course does not exist for $n > 1$). One denotes a level set of f by:

$$f^{-1}(c) = \{\boldsymbol{x} \in \mathbb{R}^n : f(\boldsymbol{x}) = c\}.$$

It perhaps helps to read this as "*the set of all inputs on which f equals c.*" For a planar function $f: \mathbb{R}^2 \to \mathbb{R}$, the level sets partition the domain into disjoint *curves* (sometimes singular - see next week for more on that); for $f: \mathbb{R}^3 \to \mathbb{R}$, the level sets are typically *surfaces*. Thinking in terms of level sets gives a more visceral approach to rates of change: you can imagine moving through the domain crossing level sets or remaining tangent to them depending on the direction of change of inputs.

GRADIENTS. The idea of a derivative as a linear transformation is fundamental, but in the more restricted setting of a scalar-valued function, an alternative to the derivative can be interpreted geometrically as a field of vectors: such *vector fields* will occupy our attention greatly in Weeks 12-14. Given a scalar field $f\colon \mathbb{R}^n \to \mathbb{R}$, there is an alternate notation and terminology for the collection of partial derivatives called the *gradient* of f:

$$\nabla f = \begin{pmatrix} \partial f/\partial x_1 \\ \vdots \\ \partial f/\partial x_n \end{pmatrix}.$$

This is interpreted as a vector at every point, in contrast to the derivative $[Df]$ which is a linear transformation (or, in the case of a scalar-valued f, a row vector) at every point. The relationship between the two is that of transpose: $\nabla f = [Df]^T$. This, then, gives a close connection with the geometry of vectors: the rate of change of the output of f when the inputs are changing at rates $\boldsymbol{h}$, is given by:

$$[Df]\boldsymbol{h} = \nabla f \cdot \boldsymbol{h}.$$

This is key to the interpretation and application of the gradient. Among all unit vectors $\boldsymbol{h}$, the dot product $\nabla f \cdot \boldsymbol{h}$ is maximized when $\boldsymbol{h}$ is aligned with ∇f. In addition, if $\boldsymbol{h}$ is tangent to a level set of f, then the rate of change of f is zero (since f is unchanging along a level set). Thus, the dot product satisfies $\nabla f \cdot \boldsymbol{h} = 0$. From these observations, we note the following important interpretations:

- ▷ *The gradient is orthogonal to the level sets of f.*
- ▷ *The gradient always points in the direction of maximal increase of f.*

This interpretation allows one to easily visualize the gradient as a field of vectors that change as one moves from point-to-point.

LINEAR APPROXIMATION. The gradient is immediately useful in determining tangent planes to implicit surfaces in 3-D. Consider a level set of $f\colon \mathbb{R}^3 \to \mathbb{R}$ and choose a point $\boldsymbol{x}_0$ on the surface. Since the gradient of f at this point is orthogonal to the level set, we can use the formula for a plane in 3-D from Weeks 1-2 to give an equation for the tangent plane to this surface: $\nabla f|_{\boldsymbol{x}_0} \cdot (\boldsymbol{x} - \boldsymbol{x}_0) = 0$. This is in contrast to the case for a parametrized surface $S\colon \mathbb{R}^2 \to \mathbb{R}^3$, in which one uses the columns of the derivative $[DS]$ as tangent vectors which span a tangent plane. Both these approaches to finding planes tangent to surfaces are the beginnings of using the derivative to perform *linear approximation.*

One additional notational approach to derivatives is at this point relevant. All of calculus operates with differential notation – the differential dx is used both in differentiation and integration. At first, one thinks of differential elements as "*infinitesimal changes*" or perhaps "*small linear changes*" in a quantity. This is not wrong, but there is more to the story. For a multivariate function f, one can use implicit differentiation to compute the differential df. This is not a derivative with respect to any particular variable: it is simply the *differential.* The formula for this looks more complicated than it really is:

$$df = \sum_{i=1}^{n} \frac{\partial f}{\partial x_i} dx_i \, .$$

When integrating in single variable calculus with $u = u(x)$, one has $du = u'(x)dx$ hardwired. Likewise, in multivariable calculus, computing df gives a combination of the differentials of the input variables. This is another approach to linear approximation: when all the partial derivatives are evaluated at a point $\boldsymbol{a}$, replacing the differentials dx_i with small changes in the x_i term combine to give the resulting approximate change in f. This is computationally no different than the matrix-vector multiplication $[Df]_{\boldsymbol{a}}\boldsymbol{h}$, where $\boldsymbol{h}$ is a vector of rates of change of inputs at $\boldsymbol{a}$, but the differential notation has some independent utility to be seen in later weeks. As a sample of what differential notation is good for, consider the problem of estimating percent changes in an output f based on percentage changes in the inputs at a point. Differentials suggest working with the following *relative rates of change* given by

$$\frac{du}{u} = d(\ln u) \, .$$

The use of the logarithm for approximating percentage changes is crucial in many corners of Statistics and Data Science. Differentials make clear the relationships between linearized percentage changes and powers, thanks to logarithm rules.

TAYLOR EXPANSION. Linear approximation is but the beginning of higher-order polynomial approximation via Taylor expansion. This is the same story as in single-variable calculus, but with more notation. For a function $f : \mathbb{R}^n \to \mathbb{R}$, the Taylor expansion of f about an input $\boldsymbol{a}$ can be written as:

$$f(\boldsymbol{x}) = \sum_I D^I f|_{\boldsymbol{a}} \frac{(\boldsymbol{x} - \boldsymbol{a})^I}{I!} \quad : \quad f(\boldsymbol{a} + \boldsymbol{h}) = \sum_I D^I f|_{\boldsymbol{a}} \frac{\boldsymbol{h}^I}{I!} \, .$$

This requires some unpacking. The *multi-index* $I = (i_1, \dots, i_n)$ is used to locate a particular monomial term in a polynomial series. For inputs $\boldsymbol{x} = (x_1, \dots, x_n)$, the monomial power $\boldsymbol{x}^I$ is the product $\boldsymbol{x}^I = x_1^{i_1} x_2^{i_2} \cdots x_n^{i_n}$. The multi-index factorial is given by $I! = i_1! \, i_2! \cdots i_n!$ (with the usual convention that $0! = 1$). The difficult part is the Ith derivative of f, $D^I f$, which is defined as follows:

$$D^I f = \frac{\partial^{i_1}}{\partial x_1^{i_1}} \frac{\partial^{i_2}}{\partial x_2^{i_2}} \cdots \frac{\partial^{i_n}}{\partial x_n^{i_n}} f \, .$$

This means that for each $k = 1 \dots n$, you take the partial of f with respect to x_k and do this I_k times (where zero derivatives means you do nothing). What saves us from an endless worry of disorder is the fact that partial differentiation operators *commute* – the order in which you take derivatives does not matter:

$$\frac{\partial^2 f}{\partial x_i \partial x_j} = \frac{\partial^2 f}{\partial x_j \partial x_i} \, .$$

Using the full formula for direct computation of Taylor series is as unpleasant as it is rare. In practice, one can chain together single-variable Taylor expansions fed with multivariate inputs – it is the Chain Rule that ensures this approach works (so long as one is careful with evaluation points).

There are a few special cases where the notation is not as imposing. When dealing with a function of two inputs, we can write out the terms in long-form:

$$f(x+h_x, y+h_y) = f(x,y) + \frac{\partial f}{\partial x}h_x + \frac{\partial f}{\partial y}h_y + \frac{1}{2}\frac{\partial^2 f}{\partial x^2}h_x^2 + \frac{\partial^2 f}{\partial x\,\partial y}h_x h_y + \frac{1}{2}\frac{\partial^2 f}{\partial y^2}h_y^2 + \cdots$$

In general, the derivatives are too many to shepherd. For low-enough orders (quadratic approximation), matrix notation once again returns. Consider the following alternate form, which has the benefit of looking more like the usual Taylor formula:

$$f(\boldsymbol{a}+\boldsymbol{h}) = f(\boldsymbol{a}) + [Df]_{\boldsymbol{a}}\boldsymbol{h} + \frac{1}{2}\boldsymbol{h}^T[D^2 f]_{\boldsymbol{a}}\boldsymbol{h} + O(|\boldsymbol{h}|^3)$$

The 1st-order term uses the derivative, as one expects. For the 2nd-order term, we can build a square matrix out of all 2nd partials, sometimes called the Hessian (but for us, simply the 2nd derivative):

$$[D^2 f]_{ij} = \frac{\partial^2 f}{\partial x_i \partial x_j}.$$

This 2nd derivative is used to define the quadratic form $Q(\boldsymbol{h}) = \boldsymbol{h}^T[D^2 f]\boldsymbol{h}$. This will give us the ability to build a 2nd-derivative test for optimization problems next week.

DISCUSSION

QUESTION 1. What are some examples of scalar fields in this room? What might their level sets look like? What about the gradient vector field?

Typical responses will be temperature. If there is a point-source light in the room (like a red dot on a smoke detector or a wireless access point), then intensity of that light (which falls off as a function of distance) gives spherical level sets and gradient fields orthogonal to those. More light-hearted answers such as "awesomeness" or "stress" can lead to interesting results.

QUESTION 2. Compute the gradient of the planar scalar fields $f = ax^2 + by^2$ for various values on constants a, b. What are the level sets of these functions?

It's a good idea to start with both constants equal to one; then both constants positive. What happens if both constants are negative? This is a good place to remind students that for many values of c, the level set $f^{-1}(c)$ may be empty, and that is ok. A mixture of negative and positive constants leads to the most interesting case of hyperbolae.

QUESTION 3. At what points in the plane are the level sets of $g = x^2 + y^2 - 2xy$ and $f = 2y - 3x$ orthogonal?

These functions are simple enough that one can draw the level sets – straight lines of slope $3/2$ in the case of f. What about the level sets of g? This at first appears to be ellipses, but $g = (x-y)^2$, so that the level sets are lines of slope 1. Oops! They are never orthogonal.

Changing to $g = x^2 + y^2 - xy$ gives an entirely different problem that requires a different solution. Use the gradients and their dot product. This is a good time to discuss the logical progression of "if the gradients are orthogonal to each other, and the level sets are orthogonal to the gradients then..."

QUESTION 4. Can you explain *why* the gradient is orthogonal to the level sets? Why does it point in the direction of maximal *increase*? How would you figure out the direction of maximal *decrease*?

All these questions are covered in the videotext, but it usually takes a second pass through for typical students to internalize what is going on (instead of memorizing the outcome). It is very worthwhile to write explicitly the relation: $[Df]\boldsymbol{h} = \nabla f \cdot \boldsymbol{h}$, to tie rates of change to geometry.

QUESTION 5. Compute tangent spaces to the following:

▷ An implicit tangent plane to $xyz - 2xy^3 + 3z^2 = 0$ at the point $(3,1,1)$.

▷ A parametrized tangent line to $\gamma(t) = \begin{pmatrix} t^2 \\ -3t \\ t^3 \end{pmatrix}$ at $t = 2$.

▷ A parametrized tangent plane to $S\begin{pmatrix} t_1 \\ t_2 \end{pmatrix} = \begin{pmatrix} t_1 + 3t_2 \\ t_1 t_2 \\ 2t_1^2 - t_2^3 \end{pmatrix}$ at $t_1 = 2, t_2 = -1$.

If students are struggling with parametrized tangent planes, try doing the two tangent lines in the previous example corresponding to the t_1 and t_2 axes.

QUESTION 6. Find the equation of a tangent hyperplane to a unit sphere in $\mathbb{R}^n$ at a given point $\boldsymbol{x}$ on that sphere.

The notation on this is cumbersome. Be sure to use the fact that $|\boldsymbol{x}| = 1$. For students who get confused, drop back to 2-D or 3-D using standard coordinate names (x, y, z) and see how it works.

QUESTION 7. For the function $f = 4x^2y^{-1/2}z^{-2}$, compute the differential df and use this to linearly approximate the value of $f(3.1, 8.7, 2.2)$.

Make sure to consider the problem of the appropriate "base point" at which to evaluate the function and its partial derivatives. If students guess at the wrong point, roll with it and compare with the better choice ex post facto.

QUESTION 8. In the previous problem, if each input can vary by as much as 1% of its value, by what percentage is the output estimated to vary, using differentials and relative rates?

Having already computed df, this can be quickly done the long way. As a follow-up (or perhaps after doing the next problem) one can go back and use logarithms to redo it quickly.

QUESTION 9. Why is it that in financial analysis and stock market data tracking, the first thing one usually does with time-series data is take its logarithm before doing any other statistics? Ask an MBA candidate in Finance and see whether you can find someone who says anything more than "*Yes, that's what you always do.*" Can you answer *why*?

This question tends to get students' attention. If they look over their notes and make a connection with $d(\ln u)$ being the relative rate of change, probe for why that matters in stock market data... What happens if you have a portfolio with multiple stocks? Are you more interested in absolute changes or percentage changes?

QUESTION 10. Given the multi-index $I = (1,3,0,2)$, what is its degree, $|I|$? What is the factorial $I!$? If $\boldsymbol{x} = (x_1, x_2, x_3, x_4)$, what is the monomial term $\boldsymbol{x}^I$?

Multi-index notation is a necessary evil, but it is algorithmic and straightforward, apart from the usual question of why $0! = 1$. If this is an issue for students, do not be afraid to ask why we should be so cavalier in defining $0!$ in this way. Among the many justifications one can pose, the best in the context of this week's material is to argue from a Taylor polynomial of a polynomial being itself.

QUESTION 11. Taylor expand the function $f(x, y) = 3 - x + 2y + 5xy - y^2$ about the point $(1,2)$ by computing partials the "long" way.

I tend to skip this: Taylor expansion the long way is very tedious. The virtue in this problem is reminding students about getting a polynomial in $(x - 1)$ and $(y - 2)$.

QUESTION 12. Taylor expand the following about the origin using composition:

- ▷ $\sinh(xy\cosh(z^2 - xy))$, including terms of degree ≤ 6
- ▷ $\ln(1 + \cos(y - xe^{xy}))$, including terms of degree ≤ 3

ASSESSMENT PROBLEMS

PROBLEM 1. Consider the function $g: \mathbb{R}^2 \to \mathbb{R}$ given by

$$g(x, y) = \frac{x^2}{4} + (y - 2)^2$$

A) Draw a picture of the level set $g^{-1}(4)$.

B) Compute the gradient ∇g and draw, on your figure from (A), the gradient vectors evaluated at several points along the level set, explaining either visually or in words the relationship between the gradients and the level set.

PROBLEM 2. Consider the function $f: \mathbb{R}^4 \to \mathbb{R}$ given by

$$f(x, y, z, t) = xy - xz^2 + \frac{y}{t} - tz$$

A) Compute the gradient ∇f and the derivative $[Df]$.

B) Explain in words what the difference is between ∇f and $[Df]$.

C) The level set $f^{-1}(0)$ passes through the point a where all inputs of f equal one. For what value(s) of C is $\boldsymbol{v} = \begin{pmatrix} 1 \\ 2 \\ 3 \\ C \end{pmatrix}$ tangent to this level set at this point a?

PROBLEM 3. Consider the level set $f^{-1}(-4)$ in $\mathbb{R}^3$ given by

$$f(x, y, z) = 3yz + xz^2 - x^3 z = -4$$

A) Find the z-coordinate(s) of the point(s) on this level set where $x = 1$, $y = 2$.

B) Give a vector that is orthogonal to this level set at a point you found in (A).

C) Using your results from (A) and (B), write down and fully simplify an equation of a plane tangent to the level set at the point.

PROBLEM 4. Let's say you know that the derivatives of functions f and g are

$$[Df] = [2x \quad -2y \quad 3] \quad : \quad [Dg] = [y \quad x \quad -2z]$$

A) Compute the gradients ∇f and ∇g.

B) Compute the gradient ∇h of the function $h = 2f - g$.

C) Give the equation of the tangent plane to the level set $f^{-1}(c)$ passing through the point $(1,2,-3)$.

PROBLEM 5. Consider the surface in $\mathbb{R}^3$ parametrized by

$$f\begin{pmatrix} t_1 \\ t_2 \end{pmatrix} = \begin{pmatrix} t_1^2\, t_2 \\ t_1\, t_2 \\ t_1\, t_2^2 \end{pmatrix}$$

A) Compute the derivative $[Df]$.

B) Give a parametrization of the tangent plane to this surface at $\begin{pmatrix} 12 \\ 6 \\ 18 \end{pmatrix} = f\begin{pmatrix} 2 \\ 3 \end{pmatrix}$, using variables s_1 and s_2 as parameters for the tangent plane.

PROBLEM 6. Consider the function $f(x, y, z) = (3x - yz)(xy - z)$.

A) Compute the gradient ∇f.

B) Write down the equation of a tangent plane to the level set $f^{-1}(15)$ at the point where $x = 1, y = 2, z = -1$.

C) Fill in the blank: *The gradient ∇f points in the direction of* ________________

PROBLEM 7. Consider the function given by

$$f(x, y, z) = \frac{1}{4}x^2 + y^2 + \frac{1}{9}z^2$$

A) Show that the level set $f^{-1}(1)$ (that is, the set of inputs where $f = 1$) contains the point $\left(\sqrt{2}, 0, \frac{3}{2}\sqrt{2}\right)$.

B) Find a vector that is orthogonal (or normal) to the level set from (A) at the point $\left(\sqrt{2}, 0, \frac{3}{2}\sqrt{2}\right)$.

C) Write down and simplify an equation of the tangent plane to the level set from (A) at the point $\left(\sqrt{2}, 0, \frac{3}{2}\sqrt{2}\right)$.

PROBLEM 8. Consider the function $f: \mathbb{R}^4 \to \mathbb{R}$ given by

$$f(x_1, x_2, x_3, x_4) = x_3^2 x_4 - x_2^3 + x_1^4$$

A) Compute the derivative $[Df]$.

B) Describe briefly in words what is meant by the level set $f^{-1}(11)$.

C) Find a vector that is orthogonal to the level set of f at the point $(0,1,2,3)$.

D) Find a vector that is tangent to the level set of f at the point $(0,1,2,3)$.

PROBLEM 9. Consider the following two scalar-valued functions on the plane:

$$h(x, y) = x^2 - 3x + 2y - y^2 + 2 \quad : \quad g(x, y) = (x + 2)(y - 1)$$

A) Compute the gradients of h and of g.

B) Complete the sentence: *the gradient of a function f points in the direction...*

C) At the origin, give a vector $\boldsymbol{v}$ such that changing the inputs of both g and h (as above) in the direction of $\boldsymbol{v}$ increases the outputs of both g and h.

D) Locate where in the plane the level sets of g and h are orthogonal to each other.

PROBLEM 10. Consider the scalar-valued function

$$G(u, v, w) = \frac{3uw^2}{v}$$

A) Compute the differential dG, using differential notation.

B) Compute and simplify as much as possible the linear approximation to the percent change in G given by dG/G.

PROBLEM 11. Consider the scalar-valued function

$$f(x,y,z) = x^2\left(\sqrt[3]{yz^2}\right) = x^2y^{1/3}z^{2/3}$$

A) Compute df using differentials.

B) Use the result of (A) to approximate to 1st order $(3.02)^2\left(\sqrt[3]{(1.04)(8.05)^2}\right)$

PROBLEM 12. Consider the scalar-valued function

$$f(x_1, x_2, x_3, x_4) = \frac{x_1^2\sqrt{x_2x_3^3}}{x_4^2}$$

A) Compute df using differentials.

B) Use the result of (A) to approximate to 1st order $\left(\frac{2.01}{1.99}\right)^2\sqrt{(4.02)(3.99)^3}$.

PROBLEM 13. Consider the function

$$f\begin{pmatrix} x \\ y \\ s \\ t \end{pmatrix} = \frac{3\sqrt[3]{s}t^2}{\sqrt{x^3y}}$$

If each input can vary by 1%, then by what percentage can the output of f vary? Use differentials to linearly approximate.

PROBLEM 14. Consider the scalar-valued function

$$F(x,y,z) = \frac{2x^3z^2}{\sqrt{y}}$$

A) Compute the differential dF, using differential notation

B) Compute and simplify as much as possible the linear approximation to the percent change in F given by dF/F.

C) If each input to F can vary by as much as 0.5%, by what percentage can the output of F vary, according to your linear approximation from part (B)?

PROBLEM 15. Compute the Taylor series about the origin of

$$f(x,y,z) = \frac{z\sin(xy)}{1-x^2-y^3} = z(\sin(xy))\frac{1}{1-x^2-y^3}$$

up to and including terms of degree seven.

PROBLEM 16. Compute the Taylor series about the origin of

$$f(x,y) = e^{y^2-x^2}\ln(1-xy)$$

up to and including terms of degree seven.

PROBLEM 17. Compute the Taylor series about the origin of

$$f(u,v,w) = 2u\ln(1+w^3-v^2) + v\cos(2\sqrt{uw})$$

up to and including terms of degree five.

PROBLEM 18. Compute the Taylor series about the origin of

$$f(x,y) = \sin(x\ln(1-xy) - 3xy)$$

up to and including terms of degree five.

PROBLEM 19. Consider the function f whose Taylor series at the origin is:

$$f(x,y,z) = 5 + x - 2y + 3x^2 + xy - \frac{1}{3}yz - z^2 + 2xy^2 - \frac{5}{7}yz^2 + x^3z + \frac{6}{7}xyz^2 + \cdots$$

A) Which terms of this series have degree three?

B) What is the gradient of f evaluated at the origin?

C) Which term in the series above corresponds to the multi-index $I = (0,1,2)$?

PROBLEM 20. A function $f(x, y)$ has this Taylor expansion at the origin:

$$3 + 2x - \frac{5}{2}x^2 + xy - \frac{3}{2}y^2 + \frac{1}{3}x^3 - \frac{1}{2}x^2y + xy^2 - \frac{2}{5}y^3 + O(|(x,y)|^4)$$

(that last part simply means "higher order terms" – don't worry about it)

A) Find the Ith derivative of f at the origin, where $I = (2,1)$ is a multi-index.

B) What is the Taylor expansion of $\partial f/\partial x$ at the origin? Of how many terms can you be confident?

ANSWERS & HINTS

PROBLEM 1. A) an ellipse at (0,2) with x-radius 1/4 and y-radius 1/2 ; B) ∇g points out away from the center of the ellipse, orthogonal to it

PROBLEM 2. A) $\nabla f = \begin{pmatrix} y - z^2 \\ x + t^{-1} \\ -2xz - t \\ -yt^{-2} - z \end{pmatrix}$, $[Df] = [y - z^2 \quad x + t^{-1} \quad -2xz - t \quad -yt^{-2} - z]$;
B) in general $\nabla f = [Df]^T$; C) where $\nabla f_a \cdot \boldsymbol{v} = 0$, namely $C = -5/2$

PROBLEM 3. A) $z = -1$ or $z = -4$; B) $\begin{pmatrix} 4 \\ -3 \\ 3 \end{pmatrix}$ or $\begin{pmatrix} 28 \\ -12 \\ -3 \end{pmatrix}$ (using the gradient at the point);
C) $4x - 3y + 3z = -5$ or $28x - 12y - 3z = 16$ respectively

PROBLEM 4. A/B) $\nabla h = 2\nabla f - \nabla g = \begin{pmatrix} 4x - y \\ -4y - x \\ 6 + 2z \end{pmatrix}$; C) $2x - 4y + 3z = -15$

PROBLEM 5. A) $[Df] = \begin{bmatrix} 2t_1t_2 & t_1^2 \\ t_2 & t_1 \\ t_2^2 & 2t_1t_2 \end{bmatrix}$; B) $G\begin{pmatrix} s_1 \\ s_2 \end{pmatrix} = \begin{pmatrix} 12 \\ 6 \\ 18 \end{pmatrix} + s_1\begin{pmatrix} 12 \\ 3 \\ 9 \end{pmatrix} + s_2\begin{pmatrix} 4 \\ 2 \\ 12 \end{pmatrix}$

PROBLEM 6. A) $\nabla f = \begin{pmatrix} 3x^{2y} + 4xy^2 + y^3 - 2xz - 2yz \\ x^3 + 4x^{2y} + 3xy^2 - 2xz - 2yz \\ -x^2 - 2xy - y^2 \end{pmatrix}$; B) $\nabla f_{1,2,-1} = \begin{pmatrix} 24 \\ 15 \\ -9 \end{pmatrix}$, so tangent plane is $24(x - 1) + 15(y - 2) - 9(z - 1) = 0$ or $24x + 15y - 9z = 45$

PROBLEM 7. B) $\nabla f = \begin{pmatrix} \sqrt{2}/2 \\ 0 \\ \sqrt{2}/3 \end{pmatrix}$; C) $\frac{\sqrt{2}}{2}(x - \sqrt{2}) + \frac{\sqrt{2}}{3}\left(z - \frac{3\sqrt{2}}{2}\right) = 0$

PROBLEM 8. A) $[4x_1^3 \quad -3x_2^2 \quad 2x_3x_4 \quad x_3^2]$; C) evaluate ∇f ; D) choose $\boldsymbol{v}$ with $\boldsymbol{v} \cdot \nabla f = 0$

PROBLEM 9. A) $\nabla h = \begin{pmatrix} 2x - 3 \\ 2 - 2y \end{pmatrix}$, $\nabla g = \begin{pmatrix} y - 1 \\ x + 2 \end{pmatrix}$; C) choose $\boldsymbol{v}$ such that $\begin{pmatrix} -3 \\ 2 \end{pmatrix} \cdot \boldsymbol{v} > 0$ and $\begin{pmatrix} -1 \\ 2 \end{pmatrix} \cdot \boldsymbol{v} > 0$; D) along the line $y = 1$

PROBLEM 10. A) $dG = 3w^2v^{-1}du - 3uv^{-2}w^2dv + 6uv^{-1}w\,dw$; B)

$$\frac{dG}{G} = \frac{du}{u} - \frac{dv}{v} + \frac{2dw}{w}$$

PROBLEM 11. A) $df = 2xy^{1/3}z^{2/3}dx + \frac{1}{3}x^2y^{-2/3}z^{2/3}dy + \frac{2}{3}x^2y^{1/3}z^{-1/3}$; B) use point $x = 3, y = 1, z = 8$ and $dx = .02, dy = .04, dz = .05$ to get approximation $36 + 0.48 + 0.48 + 0.15 = 37.11$

PROBLEM 12. B) $16.3 = 16 + 16(.01) + 2(.02) - 6(.01) + 16(.01)$ using the expansion $x_1 = 2, x_2 = 4, x_3 = 4, x_4 = 2$ and $dx_1 = .01, dx_2 = .02, dx_3 = -0.01, dx_4 = -.01$ via ; A)

$$df = 2x_1x_2^{\frac{1}{2}}x_3^{\frac{3}{2}}x_4^{-2}dx_1 + \frac{1}{2}x_1^2x_2^{-\frac{1}{2}}x_3^{\frac{3}{2}}x_4^{-2}dx_2 + \frac{3}{2}x_1^2x_2^{\frac{1}{2}}x_3^{\frac{1}{2}}x_4^{-2}dx_3 - 2x_1^2x_2^{\frac{1}{2}}x_3^{\frac{3}{2}}x_4^{-3}dx_4$$

PROBLEM 13. 13/3 % via

$$\frac{df}{f} = -\frac{3}{2}\frac{dx}{x} - \frac{1}{2}\frac{dy}{y} + \frac{1}{3}\frac{ds}{s} + 2\frac{dt}{t}$$

PROBLEM 14. A) $dF = 6x^2y^{-\frac{1}{2}}z^2\,dx - x^3y^{-\frac{3}{2}}z^2\,dy + 4x^3y^{-\frac{1}{2}}z\,dz$; C) 2.75% via ; B)

$$\frac{dF}{F} = 3\frac{dx}{x} - \frac{1}{2}\frac{dy}{y} + 2\frac{dz}{z}$$

PROBLEM 15. using the series for $\sin Z$ and $(1 + Z)^{-1}$,

$$f(x, y, z) = z\left(xy - \frac{x^3y^3}{6} + \cdots\right)(1 + (x^2 - y^3) + (x^2 - y^3)^2 + \cdots)$$
$$= xyz + x^3yz + xy^4z + x^5yz - \frac{1}{6}x^3y^3z$$

PROBLEM 16. using the series for e^Z and $\ln(1 + Z)$,

$$f(x, y) = \left(1 + y^2 - x^2 + \frac{(y^2 - x^2)^2}{2} + \frac{(y^2 - x^2)^3}{6} + \cdots\right)\left(-xy + \frac{x^2y^2}{2} - \frac{x^3y^3}{3} + \cdots\right)$$
$$= -xy + x^3y - xy^3 - \frac{x^2y^2}{2} - \frac{x^2y^4}{2} - \frac{x^4y^2}{2} - \frac{xy^5}{2} - \frac{x^5y}{2} + \frac{5x^3y^3}{6} + \cdots$$

PROBLEM 17. using the series for $\cos Z$ and $\ln(1 + Z)$,

$$f(u, v, w) = 2u\left(w^3 - v^2 - \frac{1}{2}(v^4 + \cdots)\right) + v\left(1 - uw + \frac{u^2w^2}{12} + \cdots\right)$$
$$= v - 2uvw - 2uv^2 + 2uw^3 - 2uv^4 + \frac{2}{3}u^2vw^2 + \cdots$$

PROBLEM 18. using the series for $\sin Z$ and $\ln(1 + Z)$,

$$\sin\left(x\left(-2xy + \frac{x^2y^2}{2} + \cdots\right) - 3xy\right) = -3xy - 2x^2y + \frac{1}{2}x^3y^2 + \cdots$$

PROBLEM 19. A) $2xy^2, -\frac{5}{7}yz^2$; B) $\nabla f_0 = \begin{pmatrix} 1 \\ -2 \\ 0 \end{pmatrix}$; C) $-\frac{5}{7}yz^2$

PROBLEM 20. A) $D^I f = -\frac{1}{2}I! = -1$; B) differentiate the series directly to obtain

$$\frac{\partial f}{\partial x} = 2 - 5x + y + x^2 = xy + y^2 + O(|(x, y)|^3)$$

Week 8 : Optimization

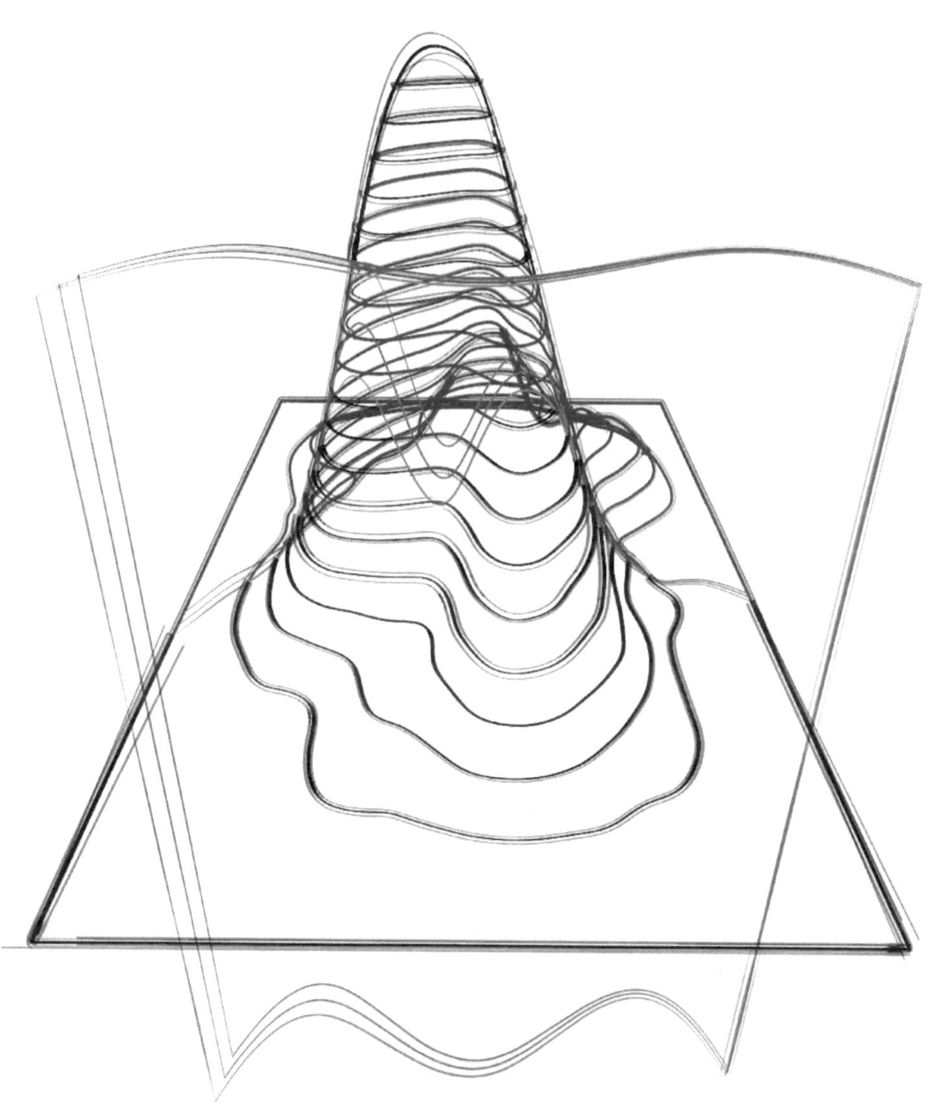

OUTLINE

MATERIALS: Calculus BLUE : Vol 2 : Chapters 14-18

TOPICS:

- ▷ Critical points and extremization of scalar-valued functions
- ▷ Classification of critical points via 2nd derivatives for planar functions
- ▷ Boundary conditions of scalar optimization problems
- ▷ Constrained optimization via substitution
- ▷ Constrained optimization via the Lagrange multiplier
- ▷ BONUS : linear regression formulae via optimization
- ▷ BONUS : Nash equilibria for symmetric 2-player games

LEARNING OBJECTIVES:

- ▷ Find critical points of scalar-valued functions
- ▷ Classify critical points in 2-D via the 2nd derivative
- ▷ Recognize saddle points in extremization problems
- ▷ Reason about boundary conditions for optimization problems
- ▷ Distinguish between local and global extrema
- ▷ Identify cost and constraint functions in constrained optimization
- ▷ Convert constrained to unconstrained optimization via parametrization
- ▷ Setup and solve the Lagrange equations for a single constraint function

PRIMER

Our work on approximating functions via Taylor expansion is about to pay off, as we generalize the max-min problems of single-variable calculus to the multivariate setting.

CRITICAL POINTS. Our first approach to optimization follows exactly the script from single-variable calculus. Consider a scalar-valued function $f: \mathbb{R}^n \to \mathbb{R}$. A *critical point* of f is any input point $\boldsymbol{a}$ at which the derivative vanishes (or is undefined). Note that a vanishing derivative means that *all* partial derivatives evaluate to zero – not just one. In addition, if the function domain is not all of $\mathbb{R}^n$ but rather some domain $D \subset \mathbb{R}^n$, then any point on the boundary of D is critical, as the derivative is technically non-existent there. With this in place, the familiar lemma holds: all extreme values of f on D must reside at critical points.

The new twist in multivariate optimization comes in the form of classifying critical points. Local minima and local maxima are as in the single-variable case; *saddle points* are critical points which are *minima* with respect to changes along some direction and *maxima* along some different direction.

A general classification scheme is not here given: such requires *eigenvalues* and other ideas from more advanced linear algebra. For this course, it suffices to remain in the 2-D case, in which everything is determined by the 2nd derivative $[D^2f]$ evaluated at the critical point. In the case of a negative determinant, the critical point is a saddle; in the case of a positive determinant, the critical point depends on the *trace* of $[D^2f]$ (the sum of the diagonal elements): if the trace is negative, the critical point is a local max; if positive, it is a local min, much like the 1-D version of the 2nd derivative test.

CONSTRAINED OPTIMIZATION. It is often the case that a quantity must be extremized subject to a constraint – a fixed amount of money, time, or space is common. Constraints also arise as a boundary condition on an unconstrained optimization problem on a domain. Whatever the cause, constrained optimization is an important subclass of problems.

Two approaches are here covered. The first involves parametrization of the constraint set and substitution into the cost function F – converting the constrained problem into an unconstrained problem on the parameters. The second approach is more novel and uses an implicit representation of the constraint set, thinking of it as the level set of some function G. The resulting *Lagrange equations* specify that the constrained extrema lie where the level sets of F and G are tangent; thus, where ∇F and ∇G are parallel. Since two vectors are parallel if they are the same up to a constant of proportionality, we have an equation – the eponymous equation of Lagrange:

$$\nabla F = \lambda \nabla G \quad \textit{or equivalently} \quad [DF] = \lambda [DG]\,.$$

This constant λ is called the *Lagrange multiplier*. A quick lemma involving the Implicit Function Theorem reveals that λ measures the rate of change of the extremal value of F with respect to the constraint value of G. (Such interpretations arise in Economics as *shadow prices*, for example.) Unfortunately, this class does not have enough bandwidth to cover how to classify the extrema found: this will have to wait until eigenvalues are taught in linear algebra.

Another topic that lies just past the bounds of this course is that of multiple constraints and multiple Lagrange multipliers. If instead of a single constraint equation, one encodes several constraints as a multi-output equation G and then constrains the optimization to the level sets $G = \boldsymbol{c}$, then the generalized Lagrange equations use a vector $\boldsymbol{\lambda}$ of Lagrange multipliers (one for each constraint):

$$[DF] = \boldsymbol{\lambda}^T [DG].$$

This is one of the less obvious benefits of using derivatives and matrix notation over gradients: it is much simpler to set up higher-dimensional optimization problems.

Both the single- and multiple-constraint versions of the Lagrange equations work by converting the constrained optimization problem to an unconstrained problem on a higher-dimensional space using the modified function

$$L = F - \boldsymbol{\lambda}^T G\,.$$

[BONUS] DATA ANALYSIS. One important application of optimization comes from basic statistics and elucidates the *best fit line* problems of elementary data analysis. Given a collection of data points of the form (x_i, y_i), the problem of finding a best fit line of the form $y = mx + b$ can be phrased as an optimization problem. The cost function

$$f(m,b) = \sum_i \big(y_i - (mx_i + b)\big)^2$$

records the net square distances from the idealized line to the data points, as measured vertically. It is a challenging but useful exercise to find the critical point of this function and classify it as a minimum. This is the beginning of the subject of *linear regression*, which extends both to multivariate data and to nonlinear representations. As the reader might guess, such generalizations are informed by Taylor series.

More generally, data analysis in the context of machine learning, AI, deep learning, and the like is highly influenced by optimization, whether it is in training a neural network, optimizing a discriminator, or other convergent phenomena. The interested reader would do well to take a dedicated course in optimization theory.

[BONUS] NASH EQUILIBRIA. Game theory provides another motivation for optimization, generally, and saddle points, specifically. In a symmetric 2-player game, each player chooses a strategy from a finite set of options (perhaps different sets between players). The game executes and, based on the strategies chosen, a certain payout from a matrix P (indexed by player strategies) is transferred from loser to winner. If each player chooses a strategy at random based on an individual probability distribution, then, over time, there is an average payout function whose inputs are the terms of the probability distributions and whose output is the average (or *expected*) payout. A *Nash equilibrium* for the game is a saddle point of this expected payout function: a choice of probability distributions which, if held, maximizes the gain of the winner (with respect to the loser changing strategies) and minimizes loss of the loser (with respect to the winner changing strategies). It is fascinating that one can analyze best-case scenarios and expected average payouts for repeated-play games using probabilities as the optimization variables.

DISCUSSION

QUESTION 1: Find and classify the critical points of the following:

- ▷ $f(x,y) = y^4 - 2xy^2 + x^3 - x$
- ▷ $g(x,y) = e^{-y}(x^2 - y^2)$

Problems like this are an excellent opportunity to discuss the hidden logical operators in solving nonlinear equations: be careful with how AND and OR operations compose.

QUESTION 2: For what values of C will the function $f(x,y) = Cx^2 + 4xy + Cy^2$ have a local max at (0,0)? Min? Saddle? Classify as a function of C.

The classification algorithm is useful, as always, but could one figure this out without using that 2nd derivative test?

QUESTION 3: Compute the second derivative matrix [*Hessian*] of the following function of six variables:

$$f(x,y,z,u,v,w) = u^3 - 3uv^2 + v^4 + w^3 - 3wx^2 + x^4 + y^3 - 3yz^2 + z^4$$

This is a good problem for conceptualizing multiple variables. As a follow-up, consider what the 2nd derivative can say about the critical point at the origin. Why is this not a surprise?

QUESTION 4: A function $f(x,y)$ has the following Taylor expansion about (0,0):

$$5 - x^2 + \frac{3}{2}xy - \frac{5}{4}y^2 + \frac{2}{9}x^3 - x^2y + \frac{2}{5}xy^2 + \frac{1}{3}y^3 + \cdots$$

Is the origin a critical point? If so, of what type?

This is a good way to recall Taylor series from the previous week, as well as to emphasize that the 2nd derivative test comes from a Taylor expansion.

QUESTION 5. [*Draw a simple closed curve in the plane*] Let's say you have a scalar-valued function on the plane constrained to this curve. What are the critical points of the function $f(x,y) = y$? What about $g(x,y) = x$? What about $h(x,y) = x - y$?

Get students to see the idea of tangent level sets, then remind them of last week's work on gradients being orthogonal to level sets. What do tangent level sets mean in terms of gradients? This is a good way to naturally bring up the Lagrange equations.

QUESTION 6: Consider the cost function $F = x^2 + y^2$ and the constraint function $G = xy$. Draw pictures of the levels sets of both functions and solve for maxima and minima graphically. Interpret the Lagrange multiplier in terms of rates of change of the optimal value.

Students may find this challenging, but it is good practice for working with level sets.

QUESTION 7: Use a Lagrange multiplier to find the critical points of the function $f(x,y,z) = x^2 + y^2 + z^2$ on the plane $ax + by + cz = 1$.

What is the geometry of this problem? Can you classify any critical points you found? How could you extend this problem from 3-D to n-D?

QUESTION 8: Use a Lagrange multiplier to re-derive the formula for the minimal distance from a point (x_0, y_0) in the plane to the line $ax + by = c$. *Hint:* minimize the square of the distance, using $f(x,y) = (x - x_0)^2 + (y - y_0)^2$ subject to the constraint of being on the line. Then, after finding this minimal distance-squared, take its square root.

Ask students if they would find it difficult to do this in arbitrary dimensions with a hyperplane.

QUESTION 9: Recall the Cobb-Douglas model of production in economics [*Week 5 Question 7*]. It says that the production P of an industrial process depends on the amount of labor L and materials M via:

$$P = \kappa L^{\alpha} M^{\beta}$$

where $\alpha + \beta = 1$. Assume that labor costs A dollars per unit, and materials cost B dollars per unit; use a Lagrange multiplier to determine how you should allocate a fixed amount of C dollars so as to maximize production.

Does your answer make sense? How do things change as worker costs increase? If there were a commodity shock of rapidly increasing prices for raw materials, how would it impact unemployment? What would you do to model two classes of labor (workers/management) and different types of raw materials?

QUESTION 10: Consider the cost functions F_1 and F_2 with potential constraint functions G_1 and G_2 where:

$$F_1 = 2x^3 + 3y^2 - 6 \quad : \quad F_2 = x^2 + y^2 - 4$$

Compare and contrast extremizing F_1 subject to $F_2 = 0$ versus extremizing F_2 subject to $F_1 = 0$. Are the extrema the same? Which problems would work better via parametrization versus via the Lagrange method? What are the advantages and disadvantages of each approach?

This is an opportunity once again to discuss parametrized versus implicit representations. The ability to classify a critical point is – at present – an advantage for the parametrized approach.

QUESTION 11: Observe that the following has a critical point at the origin:

$$f(x,y,z) = 5x^2 + 5y^2 + 9z^2 - 2xz - 2yz$$

Since it has three variables, you cannot use the 2-d method we covered in class to classify it, so what do you do?

Asking the class might yield answers such as "Try using a computer to plot something" – that's not bad. If desired, one can mention eigenvalues (to come) or Sylvester's criterion (hopefully not to come). The following approach is something that the class can do with just a little leading.

Note that f can be factored into a sum of squares. First, work with x and z:

$$f(x,y,z) = \left(5x^2 - 2xz + \frac{1}{5}z^2\right) + \frac{44}{5}z^2 - 2yz + 5y^2$$
$$= \left(\sqrt{5}x - \frac{z}{\sqrt{5}}\right)^2 + \frac{44}{5}z^2 - 2yz + 5y^2$$

If you keep going, you will have expressed f near the origin as a sum of squares, with positive coefficients in front of each square term: thus you have a minimum. This method works in general for a Hessian quadratic form near a critical point. It's not fun, but it does work!

QUESTION 12: [OPTIONAL : multiple Lagrange multipliers] Use two Lagrange multipliers to solve the following: what points on the intersection of the cylinder $x^2 + y^2 = 4$ and the plane $2x + 2y + z = 2$ are closest and farthest from the origin?

Hint: as before, you might find it helpful to maximize/minimize the square of the distance to the origin, and then take the square root of the resulting figures. For two constraints $G_1 = 0$ and $G_2 = 0$, the Lagrange equations take the form:

$$[Df] = \lambda_1[DG_1] + \lambda_2[DG_2]$$

These, together with the two constraint equations, allow one to solve for critical point(s).

QUESTION 13: What are some examples of Nash equilibria being used? In what contexts are these saddle points important?

This is a good opportunity to make connections to other disciplines and raise philosophical and moral questions. Possible tangents include:

- *The use of Nash equilibria in betting markets, evolution, warfare... what do you think?*
- *GANs [generative adversarial networks] pit a generator and discriminator neural network in a 2-player game, with the Nash equilibrium providing a convincing generator (such as thispersondoesnotexist.com). What happens when you do this with a "fake news detector" neural network?*

QUESTION 14: Compute the Nash equilibrium for the following 2-player game, where, recall, the convention is for player A to receive from player B the amount in the corresponding entry of the matrix.

$$P = \begin{bmatrix} -1 & 2 \\ 3 & -2 \end{bmatrix}$$

Compute the expected payout at the Nash equilibrium. Is there an advantage to being player A or B? In practice, it's clear that the first player has an advantage. What is not clear is how much an advantage it is and with what probability the strategies should be played.

QUESTION 15: Compute the Nash equilibrium and expected payout for a strange variant of rock-paper-scissors, where the payout matrix is given by

$$P = \begin{bmatrix} 2 & -3 & 1 \\ -3 & 5 & 0 \\ 4 & 0 & -3 \end{bmatrix}$$

For this problem, player A will choose strategies with a probability distribution $(a, b, 1-a-b)^T$ and player B with $(c, d, 1-c-d)^T$. The average payoff function will (hopefully!) have a single equilibrium: a saddle.

Finding the equilibrium is an algebraic challenge, but it uses 2-by-2 inverses and provides a nice review of skills. The resulting answer is a worthwhile surprise.

ASSESSMENT PROBLEMS

PROBLEM 1. Consider the function $f(x,y) = Cx^2 - 8xy + 2y^2$

A) For what value(s) of C does f have more than one critical point?

B) For what value(s) of C does f have a saddle point at the origin?

C) For what value(s) of C does f have a local maximum at the origin?

PROBLEM 2. Consider the function $f(x,y) = \cos x + \sin y$

A) Locate the critical points in the region where $-\frac{\pi}{2} < x < \frac{\pi}{2}$ and $-\pi < y < \pi$.

B) Compute the second derivative $[D^2 f]$.

C) Use the second derivative $[D^2 f]$ to classify the critical points from (A).

PROBLEM 3. Consider the function

$$f(x,y) = C_1x^3 + C_2xy^2 + C_3y^3 + C_4x^2 + C_5xy + C_6y^2 + C_7x + C_8y + C_9$$

where the nine coefficients C_i are all constants.

A) Compute the derivative $[Df]$ at the origin and explain: under what conditions on the constants is the origin a critical point?

B) Compute the 2nd derivative (or Hessian) $[D^2 f]$ of f at the origin. Assuming the origin is a critical point, under what conditions on the constants C_i is it a (local) minimum?

C) If $C_4 = -C_6$, what can you say about the critical point at the origin?

PROBLEM 4. Consider the function $f(x_1,x_2,x_3,x_4) = 5 + x_1^2 - 2x_2^2 + 3x_3^2 - 4x_4^2$.

A) Compute the 1st derivative $[Df]$ and 2nd derivative (i.e., Hessian) $[D^2 f]$.

B) Find all critical points of f.

C) Choose one of the critical points you found in part (B) and argue whether it should be a max, min, saddle, or degenerate point by thinking about what the function does nearby.

PROBLEM 5. Consider the function $f(x,y) = x^4y^2 - xy^3 - 8x + 8y + 13$.

A) Show that f has a critical point at $x = 1, y = 2$.

B) Compute the second derivative $[D^2 f]$.

C) Classify the critical point at (1,2).

PROBLEM 6. Find and classify the two critical points of the function

$$f(u,v) = u^3 - v^3 + uv - 7\,.$$

PROBLEM 7. Consider the function $f(x,y) = y^2 - 2y + x^2 - xy - 4x - 2$

A) Compute the first and second derivatives of f.

B) Determine the global maximum and minimum of f over the domain in the plane where $x \geq 0$.

PROBLEM 8. Consider the function $f(x,y) = x + 2xy - y^2$ on the unit disc $x^2 + y^2 \leq 1$ in the plane.

A) Find and classify all critical points in the interior of the disc (that is, ignoring the boundary).

B) Set up but do not solve the equations for finding the critical points on the boundary of this disc.

C) Will the global max be on the boundary or the interior? What about the global min?

PROBLEM 9. Consider the function $f = x^2 - y^2 + z^2$ constrained to the plane given by $x + 2y - z = 1$.

A) Use the method of Lagrange to find the critical point of this constrained f.

B) What is the value of the Lagrange multiplier λ you found in part (A)?

C) What is the value of f at the critical point you found in part (A)?

D) Do you suspect this is a max or a min or a saddle?

PROBLEM 10. Consider the function $f(x, y) = x + xy$ restricted to the circle $x^2 + y^2 = 1$.

A) Write out and simplify the Lagrange equations for this optimization problem.

B) Use the Lagrange equations to solve for the critical points. Your final answer should find three critical points on the circle.

C) Which critical point is the global maximum?

PROBLEM 11. Consider the function $f(x, y) = -3x^2 + 4xy$

A) Does f have a global maximum?

B) Find the critical point of f constrained to the line $3y - 2x = 9$.

PROBLEM 12. Consider the function $f(x, y) = 2x^2 - 6x + 5 + 2xy^2 + y^2$

A) Find and classify all the critical points of f.

B) If you constrain the function f to the line $2x - 3y = 5$ then there is a single minimum. Explain the process you would use to find that critical point.

PROBLEM 13. Consider the function $f(y, z) = 6y^{3/5}z^{2/5}$ restricted to the line $y + z = 10$ where y, z are both > 0. Your goal is to maximize f restricted to this domain.

A) Write down in full the Lagrange equations for this optimization problem.

B) Use the Lagrange method to find the unique constrained critical point.

C) Comment briefly on how you could argue that this is a maximum.

PROBLEM 14. Find and classify the extrema of the function $f(x, y) = \sqrt{3xy}$ subject to the constraint that $3x + 4y = 18$.

PROBLEM 15. Consider the function $g(x, y) = \frac{2}{3}x^3 - 2x^2y + 5xy + \frac{1}{2}y^2$

A) Compute the gradient ∇g

B) Find a point (x, y) where the gradient is zero.

C) Near this point, what does g "look like"? Is it a local max, min, etc?

D) How many nonzero terms does the Taylor series of g about the origin have?

PROBLEM 16. Consider the function $f = (x - 3y)(\cos(x^2 + y^2))(\sin(2x + y))$.

A) Taylor expand this about the origin, including only terms of degree ≤ 2.

B) This function has a critical point at the origin. Classify it.

PROBLEM 17. A) The area of an ellipse with major axis length $B > 0$ and minor axis length $C > 0$ is equal to $A = \pi BC$. Use a Lagrange multiplier to show that a circle (i.e., an ellipse where $B = C$) maximizes the area subject to the constraint that $B + C$ is a fixed number.

B) Repeat the same argument for a solid ellipsoid in 3-d with axis lengths B, C, D and volume $V = \frac{4}{3}\pi BCD$, subject to the constraint $B + C + D$ is constant: show, using the Lagrange method, that the maximum volume happens when $B = C = D$.

C) In these problems above, why is it that the critical point you found corresponds to the maximal area/volume?

PROBLEM 18. Consider the functions

$$f(x,y,z) = x^2 + 2y^2 + 3z^2 - xy - yz$$
$$g(x,y,z) = f(y,z,x) = y^2 + 2z^2 + 3x^2 - yz - xz$$

A) Compute the gradients ∇f and ∇g.

B) Write down an explicit set of equations that could be used to optimize the value of f along the surface where $g(x,y,z) = 4$.

C) Explain in words what the gradient vectors of f and g look like evaluated at a solution to the equations you wrote down in part (B).

PROBLEM 19. Assume two quantities, x and y, of items which respectively cost \$2 and \$5 per unit. Assume the total cost is \$50, and you want to minimize the function

$$F(x,y) = 2\sqrt{x}\sqrt[3]{y} = 2x^{1/2}y^{1/3}$$

A) What is the constraint equation on x and y?

B) Use the method of a Lagrange multiplier to solve for the optimal x and y. You do not need to prove that it minimizes F, but you must use the Lagrange method & show work.

PROBLEM 20. Consider the function

$$f(x,y,z) = 2z + xy - 3xz + \frac{1}{2}z^2 + xy^2 + xyz - x^2y + \frac{1}{6}z^3$$

A) Compute the gradient of f.

B) If you constrain f to the plane $z = 0$, then f has critical points at $y = 0$ and $x = 0$, or, $y = 0$ and $x = 1$. Classify these two constrained critical points.

C) Is the constrained critical point at the origin an unconstrained critical point of f?

ANSWERS & HINTS

PROBLEM 1. A) $C = 8$; B) $C < 8$; C) none: must satisfy $C > 8$ and $2C + 4 < 0$

PROBLEM 2. A) $\left(0, \pm\frac{\pi}{2}\right)$; B) $[D^2 f] = \begin{bmatrix} -\cos x & 0 \\ 0 & -\sin y \end{bmatrix}$; C) $\left(0, -\frac{\pi}{2}\right)$ max, $(0, \frac{\pi}{2})$ saddle

PROBLEM 3. A) $[Df]_0 = [C_7 \quad C_8]$; B) $[D^2 f]_0 = \begin{bmatrix} 2C_4 & C_5 \\ C_5 & 2C_6 \end{bmatrix}$, so it is a local minimum if $C_4, C_6 > 0$ and $C_5 < 2\sqrt{C_4 C_6}$; C) it is a saddle

PROBLEM 4. A) $[Df] = [2x_1 \quad -4x_2 \quad 6x_3 \quad -8x_4]$; $[D^2 f] = \begin{bmatrix} 2 & 0 & 0 & 0 \\ 0 & -4 & 0 & 0 \\ 0 & 0 & 6 & 0 \\ 0 & 0 & 0 & -8 \end{bmatrix}$; B) the origin is the only critical point and it is a saddle point

PROBLEM 5. B) $[D^2 f]_{1,2} = \begin{bmatrix} 48 & 4 \\ 4 & -10 \end{bmatrix}$; C) thus, saddle

PROBLEM 6. (0,0) is a saddle; $\left(\frac{1}{3}, -\frac{1}{3}\right)$ is a local minimum (not global)

PROBLEM 7. A) $[Df] = [2x - y - 4 \quad 2y - x - 2]$, $[D^2 f] = \begin{bmatrix} 2 & -1 \\ -1 & 2 \end{bmatrix}$; B) $\left(\frac{10}{3}, \frac{8}{3}\right)$ is the global minimum; along the edge $x = 0$, there is a local minimum at (0,1); there is no global maximum since $f \to +\infty$ as $x, y \to +\infty$

PROBLEM 8. A) $\left(-\frac{1}{2}, -\frac{1}{2}\right)$ saddle ; B) can use Lagrange or parametrize the circle ; C) since the interior critical point is a saddle, all global extrema are on the boundary

PROBLEM 9. A) $\left(-\frac{1}{2}, 1, \frac{1}{2}\right)$; B) $\lambda = -1$; C) $f = -\frac{1}{2}$; D) not a max

PROBLEM 10. A) $\begin{pmatrix} 1 + y \\ x \end{pmatrix} = \lambda \begin{pmatrix} 2x \\ 2y \end{pmatrix}$; B) $(0, -1)$ and $\left(\pm\frac{\sqrt{3}}{2}, \frac{1}{2}\right)$; C) $\left(\frac{\sqrt{3}}{2}, \frac{1}{2}\right)$ is the max

PROBLEM 11. A) nope ; B) (18,15)

PROBLEM 12. A) $\left(\frac{3}{2}, 0\right), \left(-\frac{1}{2}, \pm 2\right)$ saddles ; B) use Lagrange or parametrize

PROBLEM 13. A) $\begin{pmatrix} \frac{18}{5} y^{-\frac{2}{5}} z^{\frac{2}{5}} \\ \frac{12}{5} y^{\frac{3}{5}} z^{-\frac{3}{5}} \end{pmatrix} = \lambda \begin{pmatrix} 1 \\ 1 \end{pmatrix}$; B) $y = 6, z = 4$

PROBLEM 14. via Lagrange or parametrization, $x = 3, y = \frac{9}{4}$ and $f = \frac{9}{2}$; do not forget the endpoints at $(6, 0)$ and $\left(0, \frac{9}{2}\right)$ where $f = 0$; the interior critical point is the global max

PROBLEM 15. B) origin ; C) saddle ; D) four, of course

PROBLEM 16. A) $(x - 3y)(1)(2x + y) = 2x^2 - 5xy - 3y^2$; B) saddle

PROBLEM 17. symmetry is wonderful

PROBLEM 18. A/B) $\begin{pmatrix} 2x - y \\ 4y - x - z \\ 6z - y \end{pmatrix} = \lambda \begin{pmatrix} 6x - z \\ 2y - z \\ 4z - y - x \end{pmatrix}$; C) must be parallel

PROBLEM 19. A) $2x + 5y = 50$; B) $x = 15, y = 4$

PROBLEM 20. A) $\nabla f = \begin{pmatrix} y - 3z + y^2 + yz - 2xy \\ x + 2xy + xz - x^2 \\ 2 - 3x + z + xy + \frac{1}{2}z^2 \end{pmatrix}$; B) they are both saddles, since on this plane $[D^2 f]_{z=0} = \begin{bmatrix} -2y & 1 + 2y - 2x \\ 1 + 2y - 2x & 2x \end{bmatrix}$; C) nope

VOLUME III : INTEGRALS

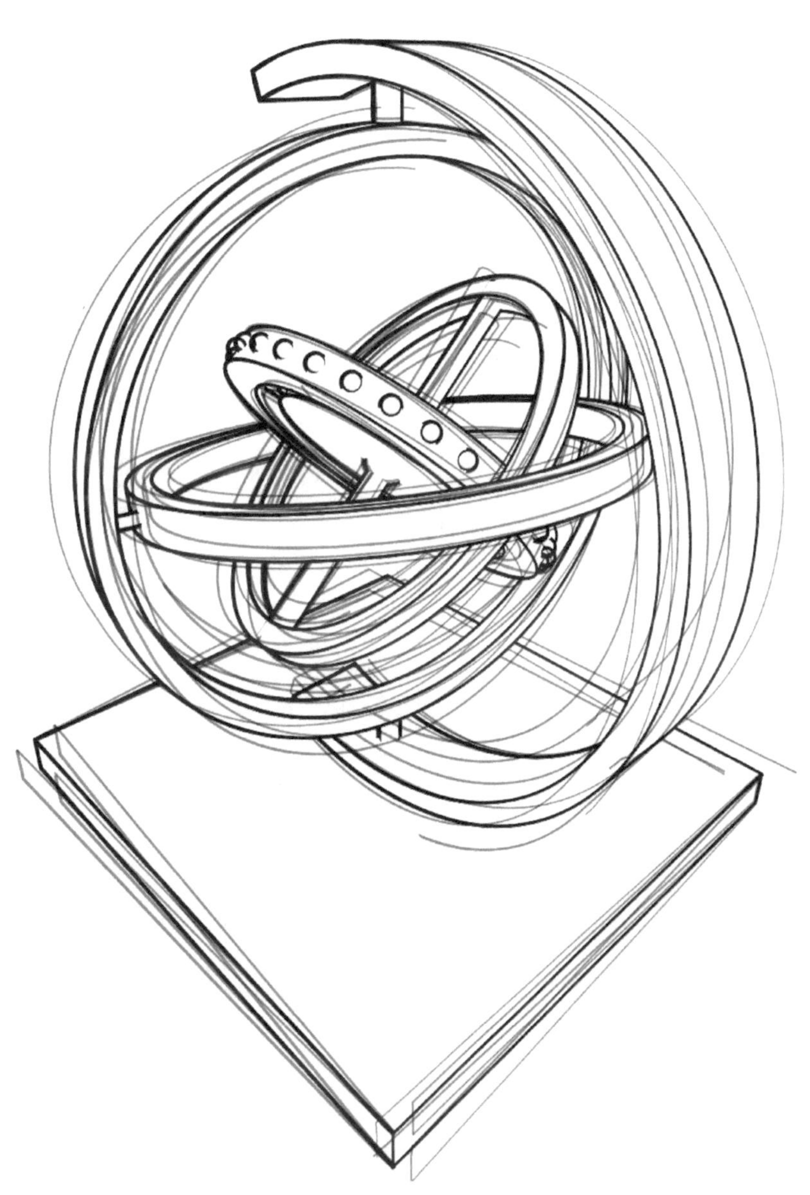

Week 9 :
Integrals & Averages

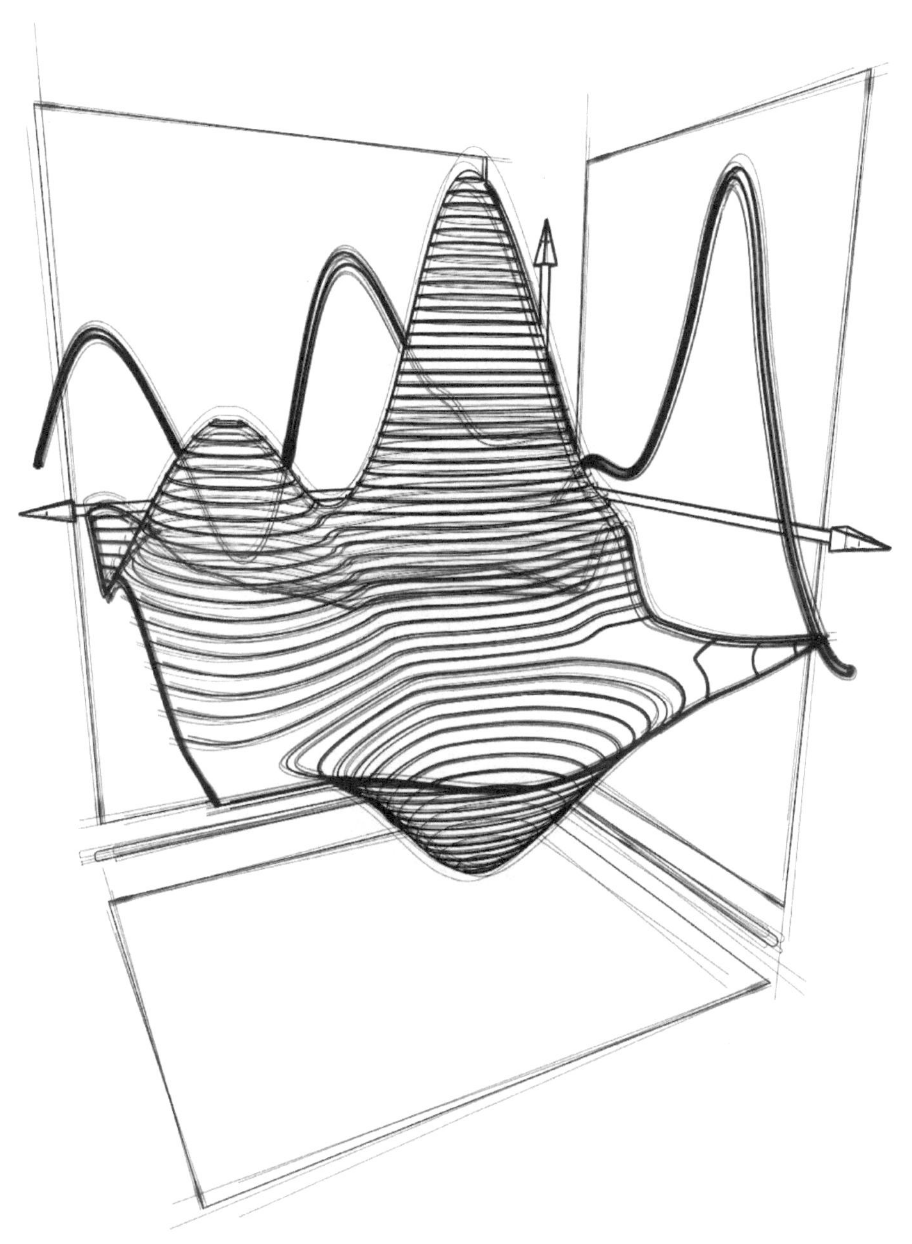

OUTLINE

MATERIALS: Calculus BLUE : Vol 3 : Chapters 1-5

TOPICS:

- ▷ Multivariate integrals as limits of Riemann sums
- ▷ Interpretation of an integrand as a signed density
- ▷ Interpretation of an integral as a signed mass
- ▷ The Fubini Theorem and iterated anti-differentiation
- ▷ Integration domains and limits for multiple integrals
- ▷ Changing order-of-integration
- ▷ Visualizing double and triple integrals
- ▷ The use of integrals for computing areas and volumes
- ▷ Averages of scalar-valued functions via integrals

LEARNING OBJECTIVES:

- ▷ Use the Fubini Theorem to evaluate multiple integrals
- ▷ Compute areas and volumes by setting up and solving integrals
- ▷ Change limits of integration under change of order of integration
- ▷ Infer planar projections of a 3-D domain of integration based on limits
- ▷ Infer limits of integration based on planar projects of the domain
- ▷ Set up and evaluate averages of functions over domains
- ▷ Estimate whether an integral is positive, negative, or zero
- ▷ Use properties of even and odd functions to simplify integrals
- ▷ Use the additivity property of integrals

PRIMER

This begins of the third quarter of our story, from differentiation to integration.

DEFINITIONS. The single-variable story of working with indefinite integrals as anti-derivatives and definite integrals as area-under-a-curve does not generalize in a straightforward fashion. More salient is the notion of a *Riemann sum* of an integrand $f: \mathbb{R}^n \to \mathbb{R}$ limited to some (reasonably nice n-dimensional) region of the domain $D \subset \mathbb{R}^n$. The mechanics of partitioning a domain and checking that a Riemann sum converges to a well-defined integral are both byzantine and unwelcome in an applications-facing course. Let us assume that every integrand continuous and bounded on a bounded domain is *integrable*, having a Riemann sum that converges as the partition element sizes shrink to zero.

INTERPRETATIONS. It is possible to compute geometric quantities by means of integrals, much as in single-variable calculus, by using an integrand of 1. In the language of differential elements, this is summing up volume elements

$$dV = dx_1 dx_2 \cdots dx_{n-1} dx_n$$

to compute n-dimensional volume. The familiar setting of $n = 1, 2, 3$ gives length, area, and volume respectively.

When integrating $f(\boldsymbol{x})dV$, one is tempted to imagine the integral as the "volume" "under" the "graph" of f. This is inconvenient. It is better to think of the domain D as a massive body with density f (which is permitted to become zero or even negative), in which case $dM = f(\boldsymbol{x})dV$ is a mass element and the integral is mass. This will be of explicit importance next week; for this week, density and mass are helpful metaphors.

FUBINI. The great coup of single variable calculus is the swap from definite to indefinite integrals flowing from the Fundamental Theorem of Integral Calculus. That theorem is as yet too lofty for us to attain in the multivariate setting (though the end of our story beckons). Our best tool at present for evaluating integrals is the Fubini Theorem which allows for iterated anti-differentiation:

$$\iint \cdots \iint f(\boldsymbol{x})\, dx_1 dx_2 \cdots dx_{n-1} dx_n = \int \left(\int \left(\cdots \int \left(\int f(\boldsymbol{x})\, dx_1 \right) dx_2 \cdots \right) dx_{n-1} \right) dx_n$$

One must exercise care with the limits of integration, but the order of integration does not change the result. (Though in practice, reordering may be advantageous for computation.)

DOUBLES & TRIPLES. An application of Fubini in $\mathbb{R}^2$ or $\mathbb{R}^3$ leads respectively to *double* or *triple* integrals. These are immediately useful for computing areas, volumes, masses, and more. For our purposes this week, there are a few challenges that will be considered carefully in the context of double and triple integrals.

- ▷ *Setting boundaries* : given a domain D, choose an order of the integration variables and set up the limits of integration to integrate over the interior of the domain.
- ▷ *Inferring geometry* : the often-harder inverse problem is to determine the geometric shape over which one is integrating given a particular ordering of variables and set limits of integration.
- ▷ *Reordering variables* : the previous problem is usually a precursor to the challenge of reordering the variables with which a multiple integral is computed.

In this course, the goals for integration are less about intricate techniques for anti-differentiating a difficult integrand (always a danger!) and more about setting up and manipulating limits of integration. It is not possible to write out a general algorithm for inferring geometry or reordering variables for a triple integral: these can be very challenging. It is best to practice with double integrals at first, then work up to handling triple integrals by ignoring the "inner" variable and working with the planar projection to the outer two variables' plane.

AVERAGES. Double and triple integrals have obvious motivations arising from geometry. Of what possible use could integration be over $\mathbb{R}^n$ be for $n > 3$? *Averages* are the simplest strongest application of higher-dimensional integrals. The same formula from single variable calculus holds here: for $f\colon \mathbb{R}^n \to \mathbb{R}$, the average of f over an n-dimensional region D in the domain is given by

$$\bar{f} = \frac{\int_D f\, dV}{\int_D dV} = \frac{1}{V} \int_D f\, dV\,,$$

where dV equals the n-dimensional volume element $dx_1 dx_2 \cdots dx_{n-1} dx_n$ and V is the volume of the region D (area in 2-D, length in 1-D, *etc.*). Next week this idea will be expanded greatly to characterize the geometry and physics of massive bodies, as well as give a quick pass over probability.

DISCUSSION

QUESTION 1: Compute the following double integral over a rectangle:

$$\int_{\pi/6}^{\pi/3}\int_{0}^{\pi/2}\cos(x-y)\ dy\, dx$$

This is a simple problem, but good for recalling the basics.

QUESTION 2: Compute the integral over the unit n-dimensional cube:

$$\int_0^1\int_0^1\cdots\int_0^1 x_1\ dx_n dx_{n-1}\cdots dx_2 dx_1$$

Encourage students to do it in two different orders to see how to use Fubini... Does the final answer make sense? Why? It is useful to look at the simplest case of $n=1$; perhaps returning to this at the end of the session, after focusing more on averages. This gives a foreshadowing of Week 10 centroids.

QUESTION 3: Reverse the order of integration of the following double integrals:

$$\int_0^4\int_{\sqrt{u}}^2 u^2+v^2\ dv\, du \quad : \quad \int_0^1\int_0^{2x} f(x,y)\ dy\, dx \quad : \quad \int_0^2\int_0^{x^2} x^2-y^2\ dy\, dx$$

What do you notice? Is there a pattern for how to change orders of integration? It is best to encourage students to use elements to think about limits: memorizing patterns does not work in general.

QUESTION 4: Set up a triple integral that computes the volume of the region that is bound by the paraboloid $x=4-y^2-z^2$ for $x\geq 0$ and by the ball of radius 2 at the origin for $x\leq 0$.

Notice how careful the wording on this is – do not say "the region between". After setting up this integral, ask how much effort it would take to solve it and compute the volume? Are there any ways to make this volume computation simpler? This emphasizes additivity of integrals.

QUESTION 5: Compute the following triple integral:

$$\int_{x=0}^{2}\int_{y=0}^{x}\int_{z=x}^{y} xyz\ dz\, dy\, dx$$

Why is the answer negative when the integrand is everywhere positive!? Have we made a mistake? Encourage students once again to begin by drawing a picture of the integration domain in the (x,y) plane. What can be said about the relative sizes of x and y? This is an opportunity to emphasize orientation and the single-variable case of reversing limits. This foreshadows the integration of 1-form fields later: the dz evaluates to a negative term since the z-coordinate is decreasing.

QUESTION 6: Compute the value of the improper integral

$$\int_0^\infty\int_0^\infty e^{-ax-by}\ dy\, dx$$

This is a good time to remind students of how improper integrals operate, as well as how the Fubini theorem operates on a product of "independent" functions over a product domain. Begin by doing this double integral "the long way"; then use Fubini to split it into the product of two cognate integrals. End with asking what happens to this if instead of a double or a triple integral one has arbitrary numbers of variables? These examples will be important in Week 10 when doing multivariate probability.

QUESTION 7: Under what conditions on the domain can one assert that:

$$\iiint \cdots \iiint f_1(x_1) f_2(x_2) \cdots f_n(x_n)\, d\mathbf{x} = \left(\int f_1\, dx_1\right)\left(\int f_2\, dx_2\right) \cdots \left(\int f_n\, dx_n\right)$$

It is worth going over why it does split in the case of a rectangular axis-aligned prism and why it does not work in general. What if you have a cube that is rotated? As always, if looking for counterexamples, 2-D is enough.

QUESTION 8: The following integral is rather difficult to compute directly, but it can be done without work if you think correctly. What is the right argument?

$$\int_{x=-1}^{1} \int_{y=0}^{\sqrt{1-x^2}} \int_{z=0}^{\sqrt{1-x^2-y^2}} (1+x)\, dz\, dy\, dx = \frac{\pi}{3}$$

This looks intimidating. The right way to start is by asking "What is the domain?" After discerning it is a quarter of a unit ball, using linearity, one can split the integral into one term than computes volume, and another that adds up the x-values. Ahha! The integral seems to evaluate to the volume...why? Use this to recall even and odd functions and how to simplify integrals based on symmetry. It is important to think of integrals in terms of Riemann sums: you can justify the cancellation of an odd integrand over a symmetric domain.

QUESTION 9: What does it mean to compute the integral of f over a single point?
Is it always zero? What additional information about f might you need?

Get students to the point where they can imagine the integral of $f: \mathbb{R}^0 \to \mathbb{R}^1$ as a Riemann sum (of sorts)., arguing for the integral as being evaluation. If students have trouble with this, give a follow-up question about the integral of $f: \mathbb{R}^2 \to \mathbb{R}^1$ over an interval (say, along the x-axis). Why must this be zero?

QUESTION 10: Compute the average of the function $f(x,y,z) = x^2 + y^2 + z^2$ over the cube defined by $-1 \leq x, y, z \leq 1$.

What kinds of symmetries are present in this problem? Would it suffice to average x^2 over the interval $0 \leq x \leq 1$? Why or why not?

QUESTION 11: Compute the average of the function $f = xy^3$ over the domain that lies outside the unit disc and inside the square $0 \leq x, y \leq 2$.

This type of problem is important, as it emphasizes additivity of integrals and weighted averages. Begin by asking how hard this would be if the domain were simply the unit disc, and whether that could be useful. This is a good problem for remembering that averages require normalization by the volume (in this case area) of the domain.

QUESTION 12: Compute the average dot product of two unit-length vectors in $\mathbb{R}^2$.

Do you expect this average to be negative, zero, or positive? Why? Does the problem become easier or harder if instead you restrict the vectors to lie in the first quadrant? The very important issue is: what is the domain of integration? If using angles as coordinates, the pair of unit vectors in the first quadrant is determined by a point in the square $[0,\pi/2] \times [0,\pi/2]$. Remember the formula for the dot product in terms of an angle between the vectors? Ahha... What would happen to the answer if instead of using angles, you average over the x-coordinates of the vectors yielding the unit square $[0,1] \times [0,1]$ and inferring the y-coordinates by the unit length constraint? Why does this not give the same answer? This is all a good foreshadowing of probability next week.

ASSESSMENT PROBLEMS

PROBLEM 1. Compute the volume of the domain in $\mathbb{R}^3$ defined by inequalities $0 \leq x \leq y^2$; $0 \leq y \leq 2$; $0 \leq z \leq y$.

PROBLEM 2. Use a triple integral to compute the volume of the domain in 3-D given by the inequalities: $0 \leq x \leq 2$: $-1 \leq y \leq 1$: $1 \leq z \leq (6 - 2x - y^3)$

PROBLEM 3. Consider the double integral

$$\int_{-1}^{8} \int_{\sqrt[3]{y}}^{2} \frac{y^2}{e^x}\, dx\, dy$$

A) Draw a careful picture of the domain of integration.

B) Reverse the order of integration (but do not evaluate the integral).

PROBLEM 4. Compute the following double integral:

$$\int_{x=0}^{2} \int_{y=x^2}^{4} \frac{x^3}{\sqrt{x^4 + y^2}}\, dy\, dx$$

PROBLEM 5. Consider the following triple integral.

$$\int_0^1 \int_{y^2}^{y} \int_{y^2}^{x^2} x^2 + y^2\, dz\, dx\, dy$$

A) Evaluate it, showing all steps.

B) Fill in the limits of integration if the order of integration is changed like so:

$$\int \int \int x^2 + y^2\, dz\, dy\, dx$$

PROBLEM 6. Consider the triple integral

$$\int_{z=0}^{2} \int_{y=0}^{\sqrt{2z}} \int_{x=2y}^{z+4} x - y\, dx\, dy\, dz$$

A) Do not evaluate the integral; rather, argue whether or not the value of this integral is positive, negative, zero, or impossible to tell (without evaluation).

B) Draw a picture of the domain of integration projected to the (y, z) plane.

C) Determine the limits of integration of this integral under the reordering:

$$\int \int \int x - y\, dx\, dz\, dy$$

PROBLEM 7. Consider the following triple integral

$$\int_{-2}^{4} \int_{z/2}^{2} \int_{-x}^{2-x^2} y - 3x^2 + 2z\, dy\, dx\, dz$$

A) Draw a picture of the domain of integration projected onto both the (x, y) and (x, z) planes.

B) Fill in the limits of integration if the order is changed like so:

$$\iiint y - 3x^2 + 2z \, dy \, dz \, dx$$

PROBLEM 8. Consider the 3-D domain whose projections to the coordinate planes are:

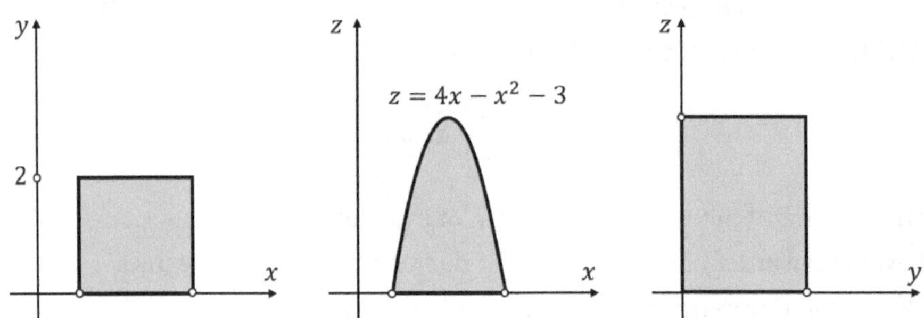

Compute the volume of this solid domain using a triple integral.

PROBLEM 9. Consider the 3-D domain whose projections to the coordinate planes are:

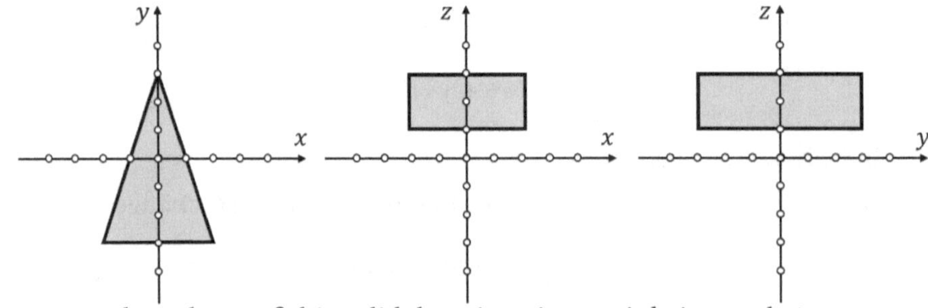

Compute the volume of this solid domain using a triple integral. Assume a unit grid in the graphs above.

PROBLEM 10. Evaluate carefully the following integral, showing all steps.

$$\int_{x_4=-1}^{1} \int_{x_3=1}^{2} \int_{x_2=0}^{x_3} \int_{x_1=x_2}^{x_3} x_1 + x_4 \, dx_1 dx_2 dx_3 dx_4$$

over the 4-dimensional cube with corners at (0,0,0,0) and (2,2,2,2).

PROBLEM 11. Compute the average of the function $f(x, y) = x^3 + 2y$ over the domain D given by $1 \leq x + y \leq 2$; $0 \leq x \leq 2$.

PROBLEM 12. Consider the following triple integral.

$$\int_{-3}^{3} \int_{0}^{\sqrt{9-y^2}} \int_{0}^{\sqrt{9-x^2-y^2}} f(x, y, z) \, dz \, dx \, dy$$

A) Describe and draw a picture of the domain of integration of the integral above.

B) If this integral evaluates to 24π, what is the average $\bar{f}$ of f over the domain?

PROBLEM 13. Consider the function $f: \mathbb{R}^4 \to \mathbb{R}$ given by

$$f(x, y, u, v) = xu - y^2 v^2$$

A) Compute and simplify the average of f over the 4-D cube where each of the four variables goes from 0 to 2.

B) Does your answer make sense?

PROBLEM 14. Consider the rectangle with corners at $(1,1), (1,0), (3,0), (3,1)$.

A) Compute the average of the function $f(x,y) = x^2 e^{-2y}$ over this rectangle.

B) Given an example of a non-constant function $g(x,y)$ that has an average of zero on this rectangle.

PROBLEM 15. Consider the 4-D domain D given by the inequalities

$$x_1^2 + x_2^2 \le 4 \quad : \quad -1 \le x_3 \le 1 \quad : \quad 0 \le x_4 \le 3$$

A) Draw pictures of D projected to the (x_1, x_2) and (x_2, x_3) planes.

B) Evaluate the integral of $f = x_1 + x_4$ over D.

PROBLEM 16. Consider the triple integral

$$\int_{x=0}^{1} \int_{y=\sqrt{1-x^2}}^{1-x} \int_{z=x}^{y} dz\, dy\, dx$$

A) Draw a careful picture of the projection of its domain onto the (x,y) plane

B) Fill in the blanks to reverse the order of integration to the following:

$$\int \int \int \quad dz\, dx\, dy$$

C) Without computing it, is the integral positive, negative, or zero?

PROBLEM 17. For a certain $f = f(x,y)$, assume you are given the following:

$$\int_{-1}^{1} \int_{x}^{1} f(x,y)\, dy\, dx = 7 \quad : \quad \int_{-1}^{1} \int_{-1}^{1} f(x,y)\, dx\, dy = 5$$

A) Draw pictures of the domains of integration of the two integrals above.

B) Determine the value of the following integral, explaining your answer.

$$\int_{-1}^{1} \int_{y}^{1} f(x,y) dx\, dy$$

PROBLEM 18. For a certain $f = f(x,y,z)$, assume you are given the following:

$$\int_{-1}^{1} \int_{-1}^{1} \int_{-1}^{y^2} f\, dz\, dy\, dx = 9 \quad : \quad \int_{-1}^{1} \int_{-1}^{1} \int_{y^2}^{1} f\, dz\, dy\, dx = 6$$

Determine the average $\bar{f}$ over the cube where x, y, z range from -1 to 1.

ANSWERS & HINTS

PROBLEM 1. the volume equals 4

PROBLEM 2. the volume equals 12

PROBLEM 3. changing order yields

$$\int_{x=-1}^{2} \int_{y=-1}^{x^3} \frac{y^2}{e^x}\, dy\, dx$$

PROBLEM 4. reverse order and substitute $u = x^4 + y^2$ to get $4(\sqrt{2} - 1)$

PROBLEM 5. A) $\frac{1}{30} - \frac{1}{55} - \frac{1}{6} + \frac{1}{7}$; B) $z = y^2 \ldots x^2$; $y = x \ldots \sqrt{x}$; $x = 0 \ldots 1$

PROBLEM 6. A) positive ; C) $x = 2y \ldots z^2 + 1$; $z = \frac{1}{2}y^2 \ldots 2$; $y = 0 \ldots 2$

PROBLEM 7.

$$\int_{x=-1}^{2}\int_{z=-2}^{2x}\int_{y=-x}^{2-x^2} y-3x^2+2z\,dy\,dz\,dx$$

PROBLEM 8. the volume equals 8/3

PROBLEM 9. the volume equals 24

PROBLEM 10. the integral evaluates to 5/2

PROBLEM 11. compute

$$\frac{1}{2}\int_0^2\int_{1-x}^{2-x} x^3+2y\,dy\,dx = 3$$

PROBLEM 12. A) this is an upper-hemisphere of radius 3 intersecting the positive z-axis ; B) $\bar{f} = 24\pi/36\pi = 2/3$

PROBLEM 13. $\bar{f} = \left(\frac{1}{16}\right)\left(-\frac{112}{9}\right) = -\frac{7}{9}$

PROBLEM 14. $\bar{f} = \left(\frac{1}{2}\right)\left(\frac{13}{3}(1-e^{-2})\right) = \frac{13}{6}(1-e^{-2})$

PROBLEM 15. A) disc of radius 2; rectangle 4×2; B) 36π

PROBLEM 16. B) $z = x \dots y\,; x = \sqrt{1-y^2} \dots 1-y\,;\ y = 0 \dots 1\,;$ C) zero, by symmetry

PROBLEM 17. B) by additivity, it equals $5-7=-2$

PROBLEM 18. the two integral domains partition the cube; compute the volumes of these domains to be 16/3 and $8 - 16/3$, then use additivity:

$$\bar{f} = \frac{1}{8}\left(9\left(\frac{16}{3}\right)+6\left(\frac{8}{3}\right)\right) = 8$$

Week 10 :
Mass & Probability

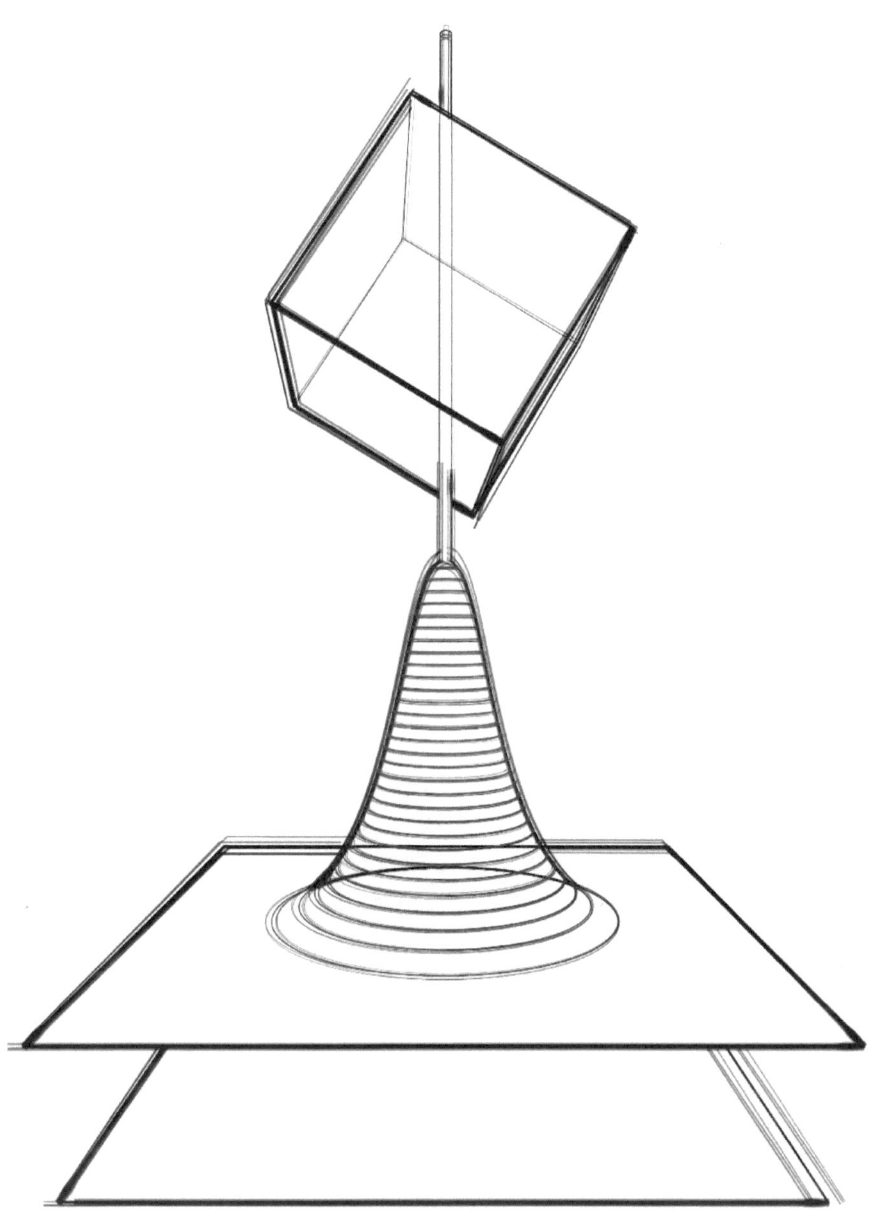

OUTLINE

MATERIALS: Calculus BLUE : Vol 3 : Chapters 6-12

TOPICS:

- ▷ Mass as the integral of a density integrand
- ▷ Centroids and centers of mass
- ▷ Moment of inertia of a 2-D or 3-D massive body about an axis
- ▷ Radius of gyration
- ▷ BONUS : The inertia matrix of a 3-D massive body; mixed moments
- ▷ BONUS : Basic solid body mechanics using the inertia matrix
- ▷ Multivariate probability density functions
- ▷ Probability as the integral of a probability density integrand
- ▷ Expectation and variance of a random variable
- ▷ Standard deviation
- ▷ Independent random variables; covariance
- ▷ Marginalization
- ▷ BONUS : Covariance matrices and their applications

LEARNING OBJECTIVES:

- ▷ Set up and compute masses, centroids, and centers of mass given densities
- ▷ Set up moment of inertia elements and compute moments of inertia
- ▷ Use the Parallel Axis Theorem to simplify moment of inertia integrals
- ▷ Compute the radius of gyration given moment of inertia and total mass
- ▷ Use and interpret multivariate probability density functions
- ▷ Set up and compute probabilities given a density
- ▷ Apply additivity of integrals to compute probabilities
- ▷ Set up expectation and variance integrals of random variables
- ▷ Compute standard deviation given variance
- ▷ Marginalize a multivariate probability density via integration & Fubini

PRIMER

This week is dedicated to applications of integrals involving mass in various forms. It serves as motivation for and an opportunity to practice computing integrals.

CENTROIDS AND CENTERS. The most common use of multivariate integrals is not the computation of area or volume but rather the computation of averages. In a geometric context, the centroid of a body $D \subset \mathbb{R}^n$ is a point $\boldsymbol{x} \in \mathbb{R}^n$ whose coordinates $\bar{\boldsymbol{x}} = (\bar{x}_1, \bar{x}_2, \dots, \bar{x}_n)$ are the average x_i coordinates over D. The student who has seen centroids covered in single variable calculus may have found it odd that the single-integral formulae for centroid coordinates $\bar{x}$ and $\bar{y}$ of the region between two graphs $y = f(x)$ and $y = g(x)$ of area A are structurally dissimilar:

$$\bar{x} = \frac{1}{A}\int_a^b f(x) - g(x)dx = \frac{1}{A}\iint x\,dA \quad : \quad \bar{y} = \frac{1}{2A}\int_a^b \left(f(x) - g(x)\right)^2 dx = \frac{1}{A}\iint y\,dA$$

The reason, of course, is that these are double integrals of x and y respectively.

The addition of a variable density ρ simply changes the volume element dV to the mass element $dM = \rho\, dV$, though it can complicate explicit computations. Centers of mass can be computed as density-weighted averages.

MOMENT OF INERTIA. There is a form of rotational mass that registers resistance to rotation about an axis in the same manner as (ordinary) mass resists linear translation via Newton's 2nd Law of Motion. This rotational mass is the *moment of inertia*, I, and it depends not only on the body but also on the axis. The moment of inertia I is the integral of *inertia element* $dI = r^2 dM$ where dM is the mass element and r is the (shortest/orthogonal) distance to the axis of rotation. In 3-D (x, y, z) coordinates, rotation about the z-axis has element $dI = (x^2 + y^2)dM$. A single point of mass M and distance R to the axis has moment of inertia $I = R^2 M$. As such, it is sometimes worthwhile to compute for a body of total mass M and total moment of inertia I the *radius of gyration*, R_g, representing the distance to the axis that would result from collapsing the body to a single point of total mass M without changing its total moment of inertia. This yields the formula $R_G = \sqrt{I/M}$.

One method of simplification is worth mentioning. If one knows the moment of inertia I_0 through the center of mass of the object and one wishes to change the axis of rotation to one that is parallel, then the integrals can be simplified. The *Parallel Axis Theorem* states that $I = I_0 + MR^2$, where R is the distance of the new axis from the old one through the center of mass.

[BONUS] INERTIA MATRIX. Changing the orientation of the axis of rotation is no mean feat via direct computation. Fortunately, linear algebra comes to the rescue. The *inertia matrix* of a body in $\mathbb{R}^3$ is a 3-by-3 matrix whose diagonal entries are the moments of inertia about the x, y, z axes respectively: denoted I_{xx}, I_{yy}, and I_{zz}:

$$[I] = \begin{bmatrix} I_{xx} & I_{xy} & I_{xz} \\ I_{yx} & I_{yy} & I_{yz} \\ I_{zx} & I_{zy} & I_{zz} \end{bmatrix} \quad : \quad I_{xy} = \int -xy\, dM \quad [etc]$$

(The term *inertia tensor* is more commonly used, but this is superfluous and intimidating.) The off-diagonal terms are called *mixed moments*, and they capture a type of asymmetry in how the mass is distributed. This symmetric matrix serves as a coefficient matrix for writing moment of inertia as a quadratic form: given any unit vector $\boldsymbol{u}$, the moment of inertia I_{uu} of the body about the axis through the center of mass in the direction of $\boldsymbol{u}$ is given by

$$I_{uu} = \boldsymbol{u}^T [I] \boldsymbol{u}\,,$$

which makes sense given the diagonal entries of $[I]$. This, together with the Parallel Axis Theorem allows one to compute all possible moments of inertia.

[BONUS] SOLID BODY MECHANICS. Many of the most confusing topics in the basic mechanics of solid bodies become much simpler with the addition of linear algebra and the inertia matrix. Consider a solid body in 3-D with inertia matrix $[I]$ rotating about some axis through its center of mass. Given a small mass element dM, let the positionplana vector $\boldsymbol{r}$ denote position relative to the center of mass, and let $\boldsymbol{v}$ be its time derivative: the velocity vector of the mass element. The following quantities encode the angular motion of the rotating body.

▷ *Angular velocity* is a vector, $\boldsymbol{\omega}$, satisfying $\boldsymbol{v} = \boldsymbol{\omega} \times \boldsymbol{r}$.
▷ *Angular momentum* is a vector, $\boldsymbol{L}$, given by $\boldsymbol{L} = [I]\boldsymbol{\omega}$.

If you are familiar with the unusual properties of angular momentum – how it jerks a spinning wheel away when the axis of rotation is changed – there is no mystery: there is only matrix-vector multiplication, with no guarantee that $\boldsymbol{L}$ must be parallel to $\boldsymbol{\omega}$. There is much more one can do with these vectors, such as defining angular *torque* $\boldsymbol{\tau}$ and angular *kinetic energy* K via:

$$\boldsymbol{\tau} = \frac{dL}{dt} = [I]\frac{d\boldsymbol{\omega}}{dt} \quad : \quad K = \frac{1}{2}\boldsymbol{\omega}^T[I]\boldsymbol{\omega}\,.$$

The inertia matrix unlocks the ability to "see" the familiar linear-motion formulae (torque is mass times acceleration; kinetic energy is one-half mass times velocity-squared) in angular form.

◊ ◊ ◊ ◊ ◊ ◊ ◊ AND NOW FOR A CHANGE IN PERSPECTIVE ◊ ◊ ◊ ◊ ◊ ◊ ◊

PROBABILITY DENSITY. Mass density appeals to physical intuition and is a clear use-case for double and triple integrals. Higher-dimensional integrals are well-suited to applications in probability. Given a domain $D \subset \mathbb{R}^n$, one speaks of choosing a point in D "at random". Whether this is done uniformly or with a bias towards certain regions of D is encoded in a probability density. A *probability density* ρ on a domain $D \subset \mathbb{R}^n$ is a nonegative scalar-valued function which, like mass density, can be integrated to obtain *probability mass*, with the condition that the total probability mass of D equals one. To determine if a point chosen at random in D lies within a subset $A \subset D$, one integrates the probability element $d\mathbb{P} = \rho\, dV$, where dV is the volume element in $\mathbb{R}^n$, like so:

$$\mathbb{P}(\boldsymbol{x} \in A) = \int_A d\mathbb{P} = \int_A \rho(\boldsymbol{x})\, dV\,.$$

One notes that $\mathbb{P}(\boldsymbol{x} \in D) = \int_D \rho = 1$ means that the odds of a randomly chosen point in D lying in D are 100%. The additivity of integration means that one can convert logical operations (*e.g.*, the probability that $\boldsymbol{x}$ lies in A and B but not in C) to an integral over a domain governed by set-theoretic operations (*e.g.*, integrate over the set $(A \cap B) - C$).

Instead of thinking in terms of picking a point out of D at random, one often reasons in terms of the individual coordinates, $x_1, \dots, x_n$, as random variables, here written with capital letters $X_1, \dots, X_n$. The probability density is then referred to as a *joint density*, relating the different random variables. This is particularly useful in applications involving multiple dependent or independent random processes. Again, one converts questions such as "*What is the probability that* $X_1 \leq X_2$" to the appropriate domain over which to integrate.

Any joint probability density on a domain $D \subset \mathbb{R}^n$ can be converted to a single-variable density via a process called *marginalization*. Via the Fubini Theorem, one can integrate out all the variables but one:

$$\rho_{X_i}(x_i) = \int_D \rho\, dx_1\, dx_2 \cdots dx_{i-1}\, dx_{i+1} \cdots dx_n\,.$$

One can check (Fubini again) that this is a probability density. This yields a notion of independence of random variables – a very useful idea in Statistics. One says that two random variables X and Y are *independent* if their joint probability density is the product of their marginalizations:

$$\rho(x,y) = \rho_X(x)\rho_Y(y) .$$

This connotes the idea that determining probabilities on the random variable X does not depend on the value of the random variable Y.

EXPECTATION AND VARIANCE : The analogy between mass and probability density continues to other features. We begin recollecting the single-variable case. The generalization of the center of mass, $\bar{x}$, to a random variable X is called the *expectation* $\mathbb{E}(X)$ or *mean*: it is simply the average of x with respect to probability density. The generalization of moment of inertia about the center of mass is called the *variance* of a random variable $\mathbb{V}(X)$, and its square root (the analogue of radius of gyration) is the familiar *standard deviation* σ_X:

$$\mathbb{E}(X) = \int x \, d\mathbb{P} \quad : \quad \mathbb{V}(X) = \int \left(x - \mathbb{E}(X)\right)^2 d\mathbb{P} \quad : \quad \sigma_X = \sqrt{\mathbb{V}(X)} .$$

All of the ideas of single-variable probability extend quickly to the multivariate case with a joint density ρ: the formulae for $\mathbb{E}, \mathbb{V}$, and σ above can be used with a marginalized probability density ρ_X. Or, as preferred, one can speak of the full expectation $\mathbb{E}$ with coordinates $\mathbb{E}(X_i)$, much as one speaks of the coordinates of a center of mass.

These ideas – besides being interesting on their own – unlock a number of avenues for exploration which must await a full probability course to unwind. In short, one can define a covariance – a degree of non-independence – of two random variables as follows:

$$\text{cov}(X,Y) = \int \left(x - \mathbb{E}(X)\right)\left(y - \mathbb{E}(Y)\right) d\mathbb{P} .$$

The covariance vanishes whenever X and Y are independent. The familiar concept of *correlation* in Statistics is a rescaled covariance: $\text{cor}(X,Y) = \text{cov}(X,Y)/\sigma_X\sigma_Y$.

[BONUS] COVARIANCE MATRIX. In the same manner that one uses an inertia matrix to encode the distribution of mass rotating about all possible axes through the center of mass, there is a matrix for keeping track of variances and covariances between all variables. The *covariance matrix* is of immense importance in data science, machine learning, statistics, and much more:

$$[\mathbb{V}] = \begin{bmatrix} \mathbb{V}(X_1) & \text{cov}(X_1,X_2) & \cdots & \text{cov}(X_1,X_n) \\ \text{cov}(X_2,X_1) & \mathbb{V}(X_2) & \cdots & \text{cov}(X_2,X_n) \\ \vdots & \vdots & \ddots & \vdots \\ \text{cov}(X_n,X_1) & \text{cov}(X_n,X_2) & \cdots & \mathbb{V}(X_n) \end{bmatrix} .$$

This matrix encodes all of the interdependencies between the random variables while enabling matrix algebra in computations. For example, when changing from one set of random variables $\boldsymbol{X} = (X_1, \dots, X_n)^T$ to another set $\boldsymbol{Y} = (Y_1, \dots Y_m)^T$ by means of a linear transformation A, so that $\boldsymbol{Y} = A\boldsymbol{X}$, the expectation and covariance transform linearly and quadratically respectively:

$$\mathbb{E}_Y = A\mathbb{E}_X \quad : \quad [\mathbb{V}_Y] = A^T[\mathbb{V}_X]A .$$

Such is very useful in data science (see the Week 11 bonus material), but it does exceed the bounds of a Calculus course.

DISCUSSION

These questions are divided into the physical & probabilistic

QUESTION 1: Consider a solid uniform-density cylinder of radius r and height h, arranged so that the centroid is at the origin and the height is aligned with the z-axis. Now imagine that you slice off the top of this object with a tilted plane [*draw figure*]. Which coordinates of the centroid of the remaining solid object *must* have changed?

This is a good visualization challenge, and a reminder of symmetry even/odd principles.

QUESTION 2: Find the centroid of the following shape: take a solid cube of side-length four centered at the origin, remove a concentric solid ball of radius two, then add a solid cylinder of radius one along the z-axis, as $-3 \leq z \leq 3$.

This is a strange problem that would be impossible to compute "by hand" without using symmetry. What if anything changes when the density switches from constant to $x^2 + y^2 + z^2$?

QUESTION 3: Compute the moment of inertia of a solid cube centered at the origin and rotated about any of the three principal axes through the centroid with density function $f = x^2 + y^2 + z^2$.

This can and should be done explicitly, noting why the choice of axis does not matter. Would things change if the axis was a skew axis through the center? It makes no difference for a constant density, as per the videos, but here?

QUESTION 4: If you know that the disc of radius R in the plane has polar inertia (rotated about the center point) $I_0 = MR^2/2$, then what else can you do? Can you compute the moment of inertia of a cylindrical shell? A solid cylinder? A solid cone?

The cone is the key example. If you think of the moment of inertia of a slice disc orthogonal to the rotation axis, then the general case of a solid of revolution about that axis reveals itself. This can lead to all sorts of interesting follow-up examples.

QUESTION 5: Here are some facts: for constant-density spherical balls (solid) and spherical shells (hollow) of radius R and total mass M, the moments of inertia are

$$I_b = \frac{2}{5}MR^2 \quad \text{versus} \quad I_s = \frac{2}{3}MR^2$$

How would you in practice compute these integrals? Try setting them up.

Begin with the solid ball. Students may be frustrated that this integral seems so hard to compute when the answer is so clean. This foreshadows coordinate changes in Week 11. As for the spherical shell, how would one compute that? It needs a surface integral – an integral with respect to the surface area element. One learned how to compute this for a surface of revolution (which this is) in single-variable calculus; but see the next problem for a better way.

Now think about this physically – the solid ball and the spherical shell. Is it harder to rotate the solid ball or the shell? Can we even compare? Are the units different? What depends on M?

QUESTION 6: Recall from the previous problem the moments of inertia for a uniform density solid ball versus a shell. Use additivity of integrals to compute this moment of inertia.

Let $\epsilon > 0$ be a small number and consider removing the ball of radius $R - \epsilon$ from the ball of radius R; computing the moment of inertia as a difference; then taking a limit as $x \to 0^+$. Wait a minute... this seems to give the wrong answer since the difference in moments of inertia is

$$\frac{2}{5}MR^2 - \frac{2}{5}M(R-\epsilon)^2 = \frac{4}{5}MR\epsilon + \frac{2}{5}M\epsilon^2$$

That cannot be right! (It is not, in fact, right: why not? Ah, the M). Replacing the mass with volume (assuming unit density) gives,

$$I_s = \frac{2}{5}\left(\frac{4}{3}\pi\right)(R^5 - (R-\epsilon)^5) = \frac{8}{3}\pi R^4\epsilon + O(\epsilon^2) = \frac{2}{3}(4\pi R^2\epsilon)R^2 + O(\epsilon^2)\,.$$

Taking the limit as $\epsilon \to 0^+$ gives the desired result, using the mass (volume) of the thin shell.

QUESTION 7: For a solid spherical ball of radius R and mass M, the moment of inertia is $I_b = 2MR^2/5$. How could you compute the moment of inertia of a hemi-spherical ball of uniform density? Does it depend on the orientation of the axis (assumed to again pass through the origin or center of the solid ball).

Sometimes it is surprising how much discussion this question can generate. Ask probing questions about what happens when you put two half-balls together and rotate that helps...

For a surprise, tell students that taking a wedge of this ball of angle θ and rotating that about the axis/edge through the center of the ball gives $I = 2MR^2/5$. This is a good chance to remind students about what happens when you normalize by mass...

QUESTION 8: Compute the moment of inertia of a unit density solid cube of side length s in $\mathbb{R}^n$ through the centroid (using an axis-aligned rotation axis, of course).

This is good practice at working out the r^2 term, as students tend to want to memorize a formula.

QUESTION 9: [OPTIONAL: inertia matrices] For an asymmetric rectangular axis-aligned prism, which axis through the centroid maximizes or minimizes the moment of inertia?

This is a difficult though great problem, since the function to be optimized is a quadratic form $Q(x) = \mathbf{x}^T[I]\mathbf{x}$, with $|\mathbf{x}| = 1$. The major and minor principal axes appear naturally. What happens with the intermediate axis? Clever students can see that it gives a natural example of a saddle point. At this point, one could lead a tangential discussion of the classic Intermediate Axis Theorem by acquiring a physical prismatic solid and rotating it in the air, explaining the relation to stability. This is far from the main storyline however.

QUESTION 10: [OPTIONAL: inertia matrices] Compute the 2-by-2 case of an inertia matrix in 2-D for a rectangular plate of width w and height h. Then use this to rotate it about the diagonal.

Ex post, one should compare to the degenerate 3-D example from the video for consistency.

QUESTION 11: [OPTIONAL: inertia matrices / angular momentum] In the case of a rectangular prism of side lengths a, b, c, what is the angular momentum if the angular velocity is $\omega = (a, b, c)^T$?

QUESTION 12: Recall the exponential probability density: $\rho = \alpha e^{-\alpha x}$ defined on the domain $D = \{x \geq 0\}$. What is its expectation and variance?

This is meant as a review of 1-D probability, which many students may need.

QUESTION 13: Say $D = \{0 \leq x \leq 3, 0 \leq y \leq 1\}$ and $\rho = C(x^2 + 2y^2)$. What value of C makes this a probability density on D? What is the probability that $x \leq y$ for a randomly chosen point (x, y)? How does this translate to a statement about the random variables X and Y? What are the integrals to compute $\mathbb{E}$ and $[\mathbb{V}]$?

This is a straightforward but important problem, as many students struggle with the basics of multivariate probability. In practice, computing the expectation and variances (especially) are too involved to do in class (or on a quiz). Setting them up is about as far as is practical except for very simple cases.

QUESTION 14: Consider the joint independent density: $\rho = \Pi_i \alpha_i e^{-\alpha_i x_i}$ on n random variables $\{X_i\}$. Say that this models a set of wait times for n people waiting in n queues. Set up (and perhaps solve if you can?) the following probabilities:

- ▷ Person 1 is done before person 2. *(In other words, $X_1 < X_2$)*
- ▷ $X_1 \leq X_2$ and all $X_i \leq 1$. *(Translate this & following into statements...)*
- ▷ $X_1 \leq X_i$ for all i. *(This is very challenging!)*
- ▷ $(X_1 \geq 1$ and $X_2 \leq X_3)$ or $(X_1 \leq 1/2$ and $X_2 \geq X_3)$ *(or similar complex conditions)*

The goal of such convoluted examples is to suggest that in probability, the logical or complex constraints can be translated into (the Boolean algebra of) geometric domains of integration. It is possible to spin off an interesting discussion of Boolean algebra/logic here.

QUESTION 15: Given a joint pdf on two variables, what is the difference between the probability $\mathbb{P}(X_1 \leq X_2)$ and the probability $\mathbb{P}(X_1 < X_2)$?

If this causes confusion, then perhaps back up to the 1-D question of what is $\mathbb{P}(X = c)$ for any constant C? Encourage students to set this up as an integral. The idea that the probability [mass] of any single point vanishes but the probability [mass] of a domain is positive is a key idea in integral Calculus.

QUESTION 16: [OPTIONAL : covariance matrices] What is the covariance matrix for the sums of three independent random variables (X, Y, Z) into sums of the form $(X + Y, Y + Z, X + Z)$?

Recall from the lectures that $[\mathbb{V}(AX)] = A[\mathbb{V}]A^T$ for a linear transformation A. What happens with more variables? Is there a recognizable pattern?

ASSESSMENT PROBLEMS

PROBLEM 1. Consider the bounded region D cut out by the graphs of $y = x^{1/3}$ and $y = x^2$.

A) Compute $\bar{y}$, the y-coordinate of the centroid of D.

B) Does your answer make sense? Why or why not?

PROBLEM 2. Consider the bounded region D in the plane given by the inequalities $0 \leq y \leq \sqrt{x}$ and $y \geq x^3 \geq 0$. If the density of this domain D is given by $\rho(x, y) = x^{-1/2}$, then compute $\bar{x}$, the x-coordinate of the center of mass of D.

PROBLEM 3. Consider the domain D in the plane defined logically by:

$$(|x| \leq 3 \;\textit{and}\; |y| \leq 5) \;\textit{but not}\; (|x| \leq 1 \;\textit{and}\; |y| \leq 3)\,.$$

This looks like a 6-x-10 rectangle at the origin with a 2-x-6 rectangle removed.

A) Compute the mass of this object if the density is $\rho(x, y) = x^2 + y^2$.

B) Where is the center of mass of this object?

PROBLEM 4. Consider a rectangle given by $0 \leq x \leq A$ and $0 \leq y \leq B$ with density $\rho(x,y) = xy$.

A) Compute the mass, M, of this object.

B) Compute the moment of inertia I of this object rotated about the y-axis.

C) Rewrite your answer to (B) using the mass: your answer should be of the form $I = CMA^2$ for some constant C. What is it?

PROBLEM 5. Consider a rectangular prism of *unit-density* that has one corner at the origin and edges along the x, y, and z axes. The length of this object along the y and z axes are L and the length along the x axis is $2L$.

A) What is the mass, M, of this object?

B) Compute the moment of inertia I of this object rotated *about the y-axis*, by setting up and solving a triple integral.

C) Rewrite your answer to (B) using the mass: your answer should be of the form $I = CML^2$ for some constant C. What is it?

D) Set up but **do not solve** an integral to compute $\overline{y}$, the y-coordinate of the centroid (or center of mass : same thing in this case).

PROBLEM 6. Fact: the moment of inertia of a flat unit-density circular disc of radius r in the (x, y) plane about the z-axis equals $I = \pi r^4/2$.

A) Use this fact to compute the moment of inertia of a solid unit-density ball of radius R at the origin, rotating about the z-axis. (*Hint: slice orthogonally to the axis of rotation and integrate...*)

B) Rewrite your answer to (A) in terms of the mass M of the solid ball.

PROBLEM 7. Recall that the moment of inertia of a constant-density ball of radius R through its center equals $I_0 = \frac{2}{5}MR^2$.

A) What is the radius of gyration R_g of this rotating solid ball?

B) Use the Parallel Axis Theorem to compute the moment of inertia I_g of this ball rotated about a parallel axis that is a distance R_g from the center – rotating the ball about the gyration axis. Write your answer as $I_g = CMR^2$ for some C.

PROBLEM 8. Consider a cube of side length L that has one corner at the origin and edges along the positive x, y, and z axes. The cube has density $\rho = x$.

A) What is the mass, M, of this cube?

B) Compute the moment of inertia I of this cube rotated about the z-axis, by setting up and solving a triple integral.

C) Compute the radius of gyration of this rotating cube, based on (A) and (B).

PROBLEM 9. Consider a unit density triangle in the plane with vertices at (0,0), (2,0), and (0,4).

A) What is the mass, M, of this triangle?

B) Compute the moment of inertia I of this triangle rotated about the y-axis.

C) Without doing more computations, would the moment of inertia of this triangle about the x-axis be greater than, less than, or the same as that about the y-axis from part (B). Explain.

PROBLEM 10. Consider a joint probability density function on two variables, X and Y, of the form $\rho(x,y) = C(x^2 + y^3)$, where the domain D is defined by $0 \leq x \leq 2$ and $0 \leq y \leq 1$.

A) What must the value of C be so that ρ is a probability density function on D?

B) What is the probability that $X \geq Y$?

C) Does your answer to part (B) make sense? Why or why not?

PROBLEM 11. Consider a joint probability density function on two random variables, X and Y, of the form $\rho(x,y) = C(x^2y + y^2)$, where the domain D is defined by $-1 \leq x \leq 1$ and $0 \leq y \leq 1$.

A) What must the value of C be so that ρ is a probability density function on D?

B) What is the probability that $X \leq 0$ and $Y \geq 1/2$?

PROBLEM 12. Consider a joint probability density function on two variables, X and Y, of the form $\rho(x,y) = C(xy)$, where the domain D is defined by $0 \leq x \leq L$ and $0 \leq y \leq L$ for some constant $L > 0$.

A) What must the value of C be so that ρ is a probability density function on D?

B) What is the probability that $X + Y \leq L$?

PROBLEM 13. Consider the domain D given by $0 \leq x \leq 1$ and $0 \leq y \leq 4$, with joint probability density on random variables X, Y on D given by

$$\rho(x,y) = \frac{9}{16}x^2\sqrt{y}$$

A) Set up and compute the integral to find $\mathbb{E}(X)$, the expected value of X.

B) Does your answer to part (A) make sense?

PROBLEM 14. Two stocks have performance modelled as random variables, X and Y, taking values in the interval $[-1,1]$. Their joint probability density is:

$$\rho(x,y) = C(2 - x^2 - y^4)$$

for $-1 \leq x \leq 1, -1 \leq y \leq 1$ and $C > 0$ some constant.

A) What must the value of the constant C be for ρ to be a probability density?

B) Compute $\mathbb{P}(X > Y)$, the probability that stock X performs better than Y.

PROBLEM 15. Consider the following joint probability density function for random variables X and Y on the domain $0 \leq x \leq 2$ and $1 \leq y < \infty$ in $\mathbb{R}^2$:

$$\rho = \frac{Cx^2}{y^3}$$

A) For what value of constant C is this a probability density?

B) Using this value of C for ρ, compute the probability that $X \leq 1$ and $Y \geq 2$.

PROBLEM 16. Consider the probability density function

$$\rho(x,y) = C(x^2 + y^2)$$

on the domain given by $0 \leq x \leq 1$ and $0 \leq y \leq 2x$.

A) For what value of constant C is this a probability density function?

B) Compute the probability that a randomly chosen point (x,y) on this domain with this probability density satisfies $x \geq \frac{1}{2}$.

C) Compute the marginal density function $\rho_Y(y)$.

PROBLEM 17. Consider a joint probability density function of the form $\rho(x,y,z) = C(x^2 + y^3 + z)$ on a domain D defined by $0 \le x \le 2,\ 0 \le y \le 1$, and $0 \le z \le 3$.

A) What must the value of C be so that ρ is a probability density function on D?

B) What is the probability that a randomly chosen point in D with respect to this probability density satisfies $x \ge y$?

C) What is the probability that $x \le y$?

PROBLEM 18. Consider a uniform probability density function $\rho(x,y) = C$, on two variables, X and Y, where the domain D is defined by $x/2 \le y \le \sqrt{2x}$.

A) What must the value of C be so that ρ is a probability density function on D?

B) Compute the marginal densities $\rho_X(x)$ and $\rho_Y(y)$. What are their domains? Can you explain why these marginalized densities are not uniform?

ANSWERS & HINTS

PROBLEM 1. $\bar{y} = \frac{12}{25}$

PROBLEM 2. $\bar{x} = \frac{7}{18}$

PROBLEM 3. A) $M = 680 - 40 = 640$; B) at the origin, thanks to symmetry

PROBLEM 4. A) $M = \frac{A^2B^2}{4}$; B) $I = \frac{A^4B^2}{8}$; C) $C = \frac{A^2}{2}$

PROBLEM 5. A) $M = 2L^3$; B) $I = \frac{10L^5}{3}$; C) $C = \frac{5}{3}$

PROBLEM 6. $I = \frac{8}{15}\pi R^5 = \frac{2}{5}MR^2$

PROBLEM 7. A) $R_g = \sqrt{\frac{2}{5}}R$; B) $I_g = \frac{2}{5}MR^2 + \frac{2}{5}MR^2 \Rightarrow C = \frac{4}{5}$

PROBLEM 8. A) $M = \frac{L^4}{2}$; B) $I = \frac{5L^6}{12}$; C) $R_g = \frac{5L}{6}$

PROBLEM 9. A) $M = 4$; B) $I = \frac{8}{3}$; C) larger I to rotate about the x-axis

PROBLEM 10. A) $C = \frac{6}{19}$; B) $\mathbb{P} = C\left(\frac{37}{12} - \frac{1}{5}\right) = C\frac{173}{60} = \frac{173}{190}$

PROBLEM 11. A) $C = 1$:-) B) $\mathbb{P} = 5/12$

PROBLEM 12. A) $C = \frac{4}{L^4}$; B) $\mathbb{P} = \frac{CL^4}{24} = \frac{1}{6}$

PROBLEM 13. $\bar{x} = \frac{3}{4}$ and $\bar{y} = \frac{12}{5}$

PROBLEM 14. A) $C = \frac{15}{88}$; B) $\mathbb{P} = C\left(\frac{44}{15}\right) = \frac{1}{2}$

PROBLEM 15. A) $C = \frac{3}{2}$; B) $\mathbb{P} = \frac{1}{128}$

PROBLEM 16. A) $C = \frac{6}{7}$; B) $\mathbb{P} = \frac{15}{16}$; C) $\rho_Y(y) = \frac{2}{7} + \frac{6}{7}y^2 - \frac{13}{28}y^3$

PROBLEM 17. A) $C = \frac{2}{37}$; B) $\mathbb{P} = \frac{154}{185}$; C) $\mathbb{P} = \frac{31}{185}$

PROBLEM 18. A) $C = \frac{3}{16}$; B) $\rho_X = \sqrt{2x} - x$ is on $[0,8]$; $\rho_Y = 2y - \frac{y^2}{2}$ is on $[0,4]$.

Week 11 :
Changing Coordinates

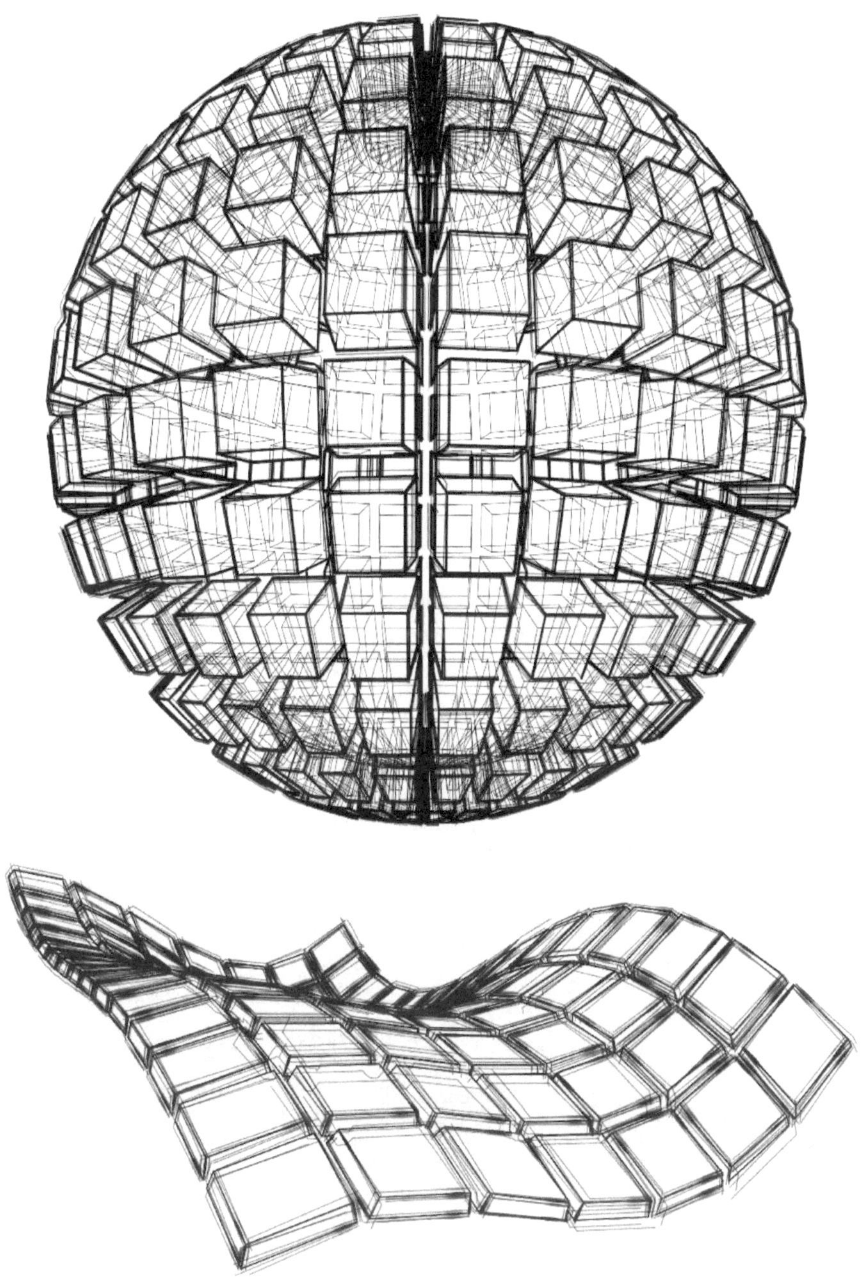

OUTLINE

MATERIALS: Calculus BLUE : Vol 3 : Chapters 13-18

TOPICS:

- ▷ Polar and cylindrical coordinates: notation, area/volume form
- ▷ Gaussians and their applications
- ▷ Spherical coordinates: notation, volume form, applications
- ▷ Arbitrary coordinate changes
- ▷ The Change of Variables Theorem and its uses
- ▷ Methods for choosing coordinates
- ▷ Surface area and surface integrals
- ▷ BONUS : High-dimensional spheres and balls
- ▷ BONUS : Gaussians and the Kalman filter in data science

LEARNING OBJECTIVES:

- ▷ Demonstrate proper use of polar/cylindrical coordinates
- ▷ Demonstrate proper use of spherical coordinates
- ▷ Apply the volume elements for cylindrical/spherical coordinates
- ▷ Distinguish when to use cylindrical versus spherical coordinates
- ▷ Use the Change of Variables Theorem to transform integrals
- ▷ Discern the proper coordinate change to transform integrals
- ▷ Compute surface area via the surface area element
- ▷ Set up and compute surface integrals for implicit/parametrized surfaces

PRIMER

Many of the applications of integrals from the previous week or two lead quickly to seemingly impossible integration problems. The goal for this week is understanding and using one integration technique that is particularly helpful. In single variable calculus, you probably called this *u-substitution*.

POLAR & CYLINDRICAL COORDINATES. For a simple example, consider the polar coordinate transformation that exchanges Euclidean (x, y) and polar (r, θ) via:

$$P\begin{pmatrix} r \\ \theta \end{pmatrix} = \begin{pmatrix} x \\ y \end{pmatrix} = \begin{pmatrix} r\cos\theta \\ r\sin\theta \end{pmatrix}$$

What does this do to area? Perhaps you recall tricks in single-variable calculus. The better approach is direct conversion of the area element: $dA = dx\,dy = r\,dr\,d\theta$. Where does the r come from? Often, units are invoked to justify the additional r paired with $d\theta$. Examining the geometry of polar coordinates also helps. In 3-D, adding the usual z-axis yields cylindrical coordinates, with corresponding volume element $dV = dx\,dy\,dz = r\,dr\,d\theta\,dz$. For domains that are well-suited to either of these coordinate systems, integrals can be done much more simply.

SPHERICAL COORDINATES. Cylindrical coordinates are not the only useful system available. Spherical coordinates come in competing notational variants: in this course: ρ is used for the radial coordinate; θ is the usual polar angle from $0 \dots 2\pi$; and $0 \leq \phi \leq \pi$ is the angle from the positive z-axis. The corresponding spherical coordinate transformation is:

$$S\begin{pmatrix}\rho\\ \theta\\ \phi\end{pmatrix} = \begin{pmatrix}x\\ y\\ z\end{pmatrix} = \begin{pmatrix}\rho\cos\theta\sin\phi\\ \rho\sin\theta\sin\phi\\ \rho\cos\phi\end{pmatrix}$$

What does this do to volume? This is not so obvious. In spherical coordinates, the volume element transforms as $dV = dx\,dy\,dz = \rho^2 \sin\phi\, d\rho\, d\phi\, d\theta$. Arguing via geometry is much more difficult; arguing via units explains only the ρ^2 term. What one cannot argue with is how singularly effective spherical coordinates can be at solving very difficult integration problems adapted to spherical geometry.

CHANGE of VARIABLES THEOREM. The answer to what happens in cylindrical and spherical coordinates comes from a fundamental result. Given a change of coordinates on $\mathbb{R}^n$ from $\boldsymbol{u} = (u_1, \dots, u_n)$ to $\boldsymbol{x} = (x_1, \dots, x_n)$ via $\boldsymbol{u} = F(\boldsymbol{x})$, the volume element transforms as

$$d\boldsymbol{u} = du_1\, du_2 \cdots du_n = |\det[DF]|\, dx_1 dx_2 \dots dx_n = |\det[DF]|\, d\boldsymbol{x}$$

This is the essential part of the Change of Variables Theorem, and it is the true multivariate version of the "u-subs" method learned in elementary integration.

After verifying what happens with cylindrical and spherical coordinates, this deep result is worth contemplation. Recall from Week 4 how we interpreted the determinant as the change-in-volume induced by a linear transformation? Recall from Weeks 5-7 how the derivative $[DF]$ is a linear transformation and the linear approximation to the nonlinear F? The Change of Variables Theorem ties together so much of what has been learned this semester to give an integration technique for difficult problems.

SURFACE INTEGRALS. There are several instances in which one might want to compute surface area of a curved surface in 3-D. Instead of a volume element, there is a surface area element which can be integrated. For example, based on what we know of the volume element in spherical coordinates, we can fix a sphere of radius R and reduce the spherical volume element to $R^2 \sin\phi\ d\phi\ d\theta$.

Given a parametrized surface defined by $G: \mathbb{R}^2 \to \mathbb{R}^3$, the two column vectors of the derivative $[DG]$ span an infinitesimal parallelogram (the *surface area element* $d\sigma$) on the tangent plane to the surface. Recall from Week 2 that the area of a parallelogram in 3-D equals the length of the cross product of the spanning vectors. This gives an effective formula for the surface area element:

$$d\sigma = \left|\frac{\partial G}{\partial s} \times \frac{\partial G}{\partial t}\right| ds\, dt\,.$$

In the case of an implicitly defined surface of the form $z = z(x, y)$, one shows that the above formula specializes to

$$d\sigma = \sqrt{1 + \left(\frac{\partial z}{\partial x}\right)^2 + \left(\frac{\partial z}{\partial y}\right)^2}\ dx\, dy\,.$$

The surface area element can be integrated to compute surface area; or it can be used in *surface integrals* to compute centroids, moments, and other features of surfaces in 3-D.

Both formulae above are unsatisfying, as they are explicitly three-dimensional, relying on constructs like the cross product. A much more general formula for the surface area element exists for $G: \mathbb{R}^2 \to \mathbb{R}^n$ which reveals the deep connection to the Change of Variables Theorem:

$$d\sigma = \sqrt{\det([DG]^T[DG])}\ ds\,dt\,.$$

[BONUS] GAUSSIANS & DATA FUSION. The most iconic probability density is a *Gaussian*. The standard zero-mean Gaussian on n variables with unit variances and all variables independent (pairwise covariances vanish) is given by

$$g(\boldsymbol{x}) = \frac{1}{\sqrt{(2\pi)^n}} e^{-\frac{1}{2}\boldsymbol{x}\cdot\boldsymbol{x}}\,,$$

which one can show to be of unit total mass on $\mathbb{R}^n$ by a combination of Fubini and polar coordinates in the $n = 2$ case. It is easy and elegant to change the mean to $\mathbb{E} \in \mathbb{R}^n$ and all the variances and covariances by using the covariance matrix $[\mathbb{V}]$ from Week 10, obtaining the following general Gaussian:

$$g(\boldsymbol{x}) = \frac{1}{\sqrt{(2\pi)^n \det[\mathbb{V}]}} e^{-\frac{1}{2}(\boldsymbol{x}-\mathbb{E})^T[\mathbb{V}]^{-1}(\boldsymbol{x}-\mathbb{E})}\,.$$

This is not a formula one memorizes, but general Gaussians are central to modern applications in data science. Consider, *e.g.*, data filtering and fusion, in which one estimates n variables $x_1, x_2, \dots, x_n$ with a temporal evolution model and updates at various times. Instead of trying to track the exact values of the x_i, one estimates the probability density of these random variables. Using a Gaussian is very efficient, as one need only keep track of $\mathbb{E}$ and $[\mathbb{V}]$. A classic example of data fusion is the *Kalman filter*, which consists of three steps: (1) a model step, which predicts the next state based on the present state, $\boldsymbol{y} = F(\boldsymbol{x})$; (2) a measurement step, which estimates the mean and covariance matrix; and (3) a fusion step, which takes the predicted and measured densities and multiplies them together. Using the fact that the product of two general Gaussians is again (up to rescaling) a generalized Gaussian, one has the following formula for the Kalman filter acting on initial mean and covariance $(\mathbb{E}_0\,, [\mathbb{V}_0])$ and measured mean and covariance $(\mathbb{E}_m\,, [\mathbb{V}_m])$:

$$\text{predicted} : \mathbb{E}_p = F(\mathbb{E}_0) : [\mathbb{V}_p] = [DF][\mathbb{V}_0][DF]^T$$

$$\text{fused mean} : \mathbb{E}_f = [\mathbb{V}_m]([\mathbb{V}_p] + [\mathbb{V}_m])^{-1}\mathbb{E}_p + [\mathbb{V}_m]([\mathbb{V}_p] + [\mathbb{V}_m])^{-1}\mathbb{E}_m$$

$$\text{fused covariance} : [\mathbb{V}_f] = [\mathbb{V}_m]([\mathbb{V}_p] + [\mathbb{V}_m])^{-1}[\mathbb{V}_p]$$

This looks complicated, but it is only a little beyond the bounds of this course, and it is very much at the heart of modern applications in control, estimation, and modelling.

[BONUS] THE GEOMETRY of DATA. Most students know the statistics of the 1-D Gaussian: the mass within 1, 2, and 3 standard deviations is ~68%, ~95%, and >99% respectively. This is *not* the case for a standard higher-dimensional Gaussian, as can be seen with a little effort in 2-D using integration and polar coordinates. Working in higher dimensions requires a higher-dimensional version of spherical coordinates, which has its own uses and fascinations. The answer to "*Where is the mass in a Gaussian?*" takes an extensive detour into the geometry of balls and spheres. With a lot of detailed work involving high-dimensional spherical coordinates and the classic gamma function $\Gamma(z)$ from single-variable calculus, we have the following formulae for the volume of the radius R ball $B_n(R)$ and the surface volume of its boundary sphere $S_n(R)$ in $\mathbb{R}^n$:

$$\mathrm{vol}_n(B_n(R)) = \frac{2\pi^{\frac{n}{2}}R^n}{n\Gamma\left(\frac{n}{2}\right)} \quad : \quad \mathrm{vol}_{n-1}(S_n(R)) = \frac{2\pi^{\frac{n}{2}}R^{n-1}}{\Gamma\left(\frac{n}{2}\right)}.$$

What is interesting about this is that the gamma function – recall it is a proxy for the factorial – grows very rapidly, so that for any fixed radius, the volumes are going to zero rapidly in dimension. This means that, in the context of statistics, the probability of being within distance R of the mean of a standard Gaussian goes rapidly to zero as the dimension increases. No matter how many standard deviations away from the mean one looks, there is hardly any mass there in sufficiently large dimensions.

This is a disquieting result, one of many such in high-dimensional data. Resolving this paradox must await a more advanced course, in which one can prove that a unit-variance Gaussian in dimension n has most of its mass accumulated near a sphere about the mean of radius $\sqrt{n}$.

DISCUSSION

This week's questions should echo mass/probability concepts...

QUESTION 1: Where is the centroid of a solid uniform-density hemisphere?

This is a good opening question. Get students to think in terms of the right coordinate system. Where should the origin be? Where should the axes be situated? What does symmetry tell you?

QUESTION 2: Consider the uniform probability density on the unit hemisphere with $z \geq 0$. What is this density? What is the expectation $\mathbb{E}$ of the vector of random variables (X, Y, Z)? (If Problem 1 has been done, then this is worth asking!) Think about the expectation of the variable Z, $\mathbb{E}(Z)$, then, consider what happens with the expected valued of the spherical coordinates. Of course, $\mathbb{E}(\rho)$ makes sense, but what is it? Is it the same as $\mathbb{E}(Z)$?

Students may guess that since the centroid is along the z-axis where $\phi = 0$, the $\mathbb{E}(\rho) = \mathbb{E}(Z)$; however, this does not hold, as can be seen by comparing the integrals of $\sin\phi \, d\phi$ and $\cos\phi \sin\phi \, d\phi$. This is a good chance to rethink level sets where z is constant (which have the largest mass?) and where ρ is constant (now, which have the largest mass?). One can and should spend a long time on this problem.

QUESTION 3: Derive the results from last week's discussion about the moment of inertia of a uniform density ball/shell of radius R rotated about the centroid:

$$I_b = \frac{2}{5}MR^2 \quad : \quad I_s = \frac{2}{3}MR^2$$

That is by no means an easy problem – the integrals require remembering some trig formulae. However, it is clear that setting up these integrals using spherical coordinates is helpful.

QUESTION 4: Consider the probability density $f = C(x^2 + y^2)$ on a domain D given by a solid cylinder of unit radius about the z-axis for $-1 \leq z \leq 1$. Where is the mean $\mathbb{E}$? Compute the variance $\mathbb{V}$ of this joint pdf.

Symmetry should be used to place the mean at the origin. For the variance, this is a nice integral in cylindrical coordinates.

QUESTION 5: Consider a cube in 3-D centered at the origin whose side length is 2 and which has the following as its density function:

$$f(x,y,z) = \frac{1}{\left(\sqrt{x^2+y^2+z^2}\right)^{\alpha}}$$

for some constant $\alpha > 0$. This is rather dangerous, as the density becomes infinite at the center of the cube. For which values of α does the cube have a total mass that is finite?

This is a very good discussion question, since it suggests the use of spherical coordinates for the density, but the limits of integration on the cube are awful in spherical coordinates. Try to get the students to come to the realization that the mass outside a small ball is finite; thus, one can compute the mass of a ball about the origin. This then becomes a good way of seeing how critical the volume element is in dealing with this infinite density.

QUESTION 6: The standard Gaussian in 2-D is a product of standard Gaussians in 1-D; thus, via Fubini:

$$\iint_{\mathbb{R}^2} \frac{1}{2\pi} e^{-(x^2+y^2)}\, dA = \left(\int_{\mathbb{R}} \frac{1}{\sqrt{2\pi}} e^{-x^2}\, dx\right)\left(\int_{\mathbb{R}} \frac{1}{\sqrt{2\pi}} e^{-y^2}\, dy\right) = 1$$

Use polar coordinates to compute the mass of a 2-D Gaussian within 1, 2, and 3 standard deviations of the mean. What domains of integration are these?

This is covered in the bonus videos on Gaussians but is very valuable to do live. Be sure to remind students of the oft-memorized 68-95-99.7 rule from statistics. Students will need to use a computer to get the estimated values of the double integral over discs of radius 1, 2, and 3. For students who are curious about what happens in higher dimensions, please refer them to the bonus videos.

QUESTION 7: What coordinate change would you use to evaluate

$$\iint xy\,(x^2+y^2)\, dx\, dy$$

over the domain given by $1 \le xy \le 4$ and $1 \le x^2 - y^2 \le 3$? An alternative option is

$$\iint x^2y^2\,(y^2-x^2)\, dx\, dy$$

over the domain given by $1 \le xy \le 4$ and $1 \le y - x \le 3$.

Discuss that sometimes the coordinate transformation presents itself; other times, not so much. This problem warns against trying too hard to simplify the integrand before computing how the volume element changes. (Students may wonder whether there's a typo with the integrand...)

QUESTION 8: Use a change of coordinates to evaluate the challenging integral

$$\int_{y=0}^{1}\int_{x=0}^{1-y} e^{(x+y)^2}\, dx\, dy$$

This presents several difficulties – the choice of $u = x + y$ and $v = y$ should not be too hard to see. However, the transformed integrand appears impossible and requires changing the order from $du\, dv$ to $dv\, du$. The other difficulty in this problem is the limits of integration, which are not rectangular. If $u = x + y$ then the limits on u are from $-v$ to 1. This is challenging.

QUESTION 9: The following coordinate change gives "toroidal" coordinates about a circle of radius 3 in the (x, y) plane, with a longitudinal angle about the circle, ψ, and cross-sectional polar coordinates (r, ϕ):

$$\begin{pmatrix} x \\ y \\ z \end{pmatrix} = \begin{pmatrix} (3 + r\cos\phi)\cos\psi \\ (3 + r\cos\phi)\sin\psi \\ r\sin\phi \end{pmatrix}$$

What is the resulting volume element?

This is not hard to set up; however, the determinant computation is very involved, and simplifying the resulting volume element is a challenge. This is best left for the most eager students to pursue.

QUESTION 10: What is the total charge on sphere when surface charge density is of the form κz^2 for a constant κ?

Remember to start with the spherical volume element $dV = \rho^2 \sin\phi \; d\rho \, d\phi \, d\theta$ and reduce to the surface area element $d\sigma = R^2 \, d\phi \, d\theta$, spending lots of time on this step if needed. Students often struggle with the surface area element – cylindrical and spherical coordinates are excellent special cases that can assist with intuition.

QUESTION 11: Compute $d\sigma$ for a torus of major radius 3 and minor radius 1, using the surface parametrization from Question 9:

$$\begin{pmatrix} x \\ y \\ z \end{pmatrix} = \begin{pmatrix} (3 + \cos\phi)\cos\psi \\ (3 + \cos\phi)\sin\psi \\ \sin\phi \end{pmatrix}$$

This is, as with most surface area elements, ugly, but with some decent simplifications possible.

QUESTION 12: Integrate the function $f = z(x^2 + y^2)$ over the parametrized surface given by $x = u\cos v \; ; \; y = u\sin v \; ; \; z = u \; ; 0 \leq u, v \leq 1$.

QUESTION 13: Compute the surface area of the cone $z = \sqrt{x^2 + y^2}$ for $z \leq 4$.

These types of surface integral problems are usually difficult to solve explicitly. At least one implicit and one parametrized problem should be practiced.

ASSESSMENT PROBLEMS

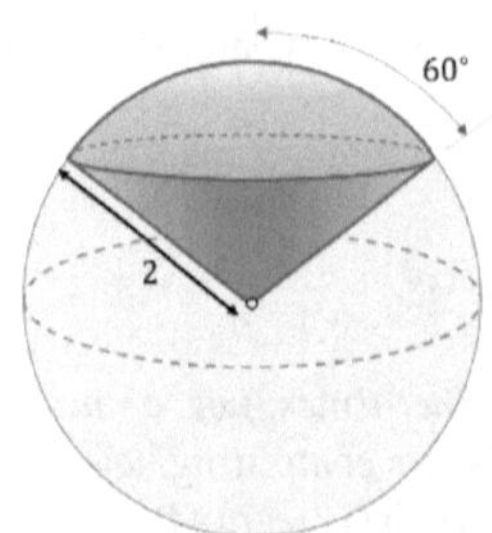

PROBLEM 1. Use spherical coordinates to compute the volume of the solid conical domain as shown.

PROBLEM 2. Compute the volume of the solid D given in spherical coordinates by:

$$0 \leq \rho \leq 1 + \cos\phi \quad : \quad 0 \leq \phi \leq \pi \quad : \quad 0 \leq \theta \leq \pi/2$$

PROBLEM 3. Convert the following integral to cylindrical coordinates:

$$\int_{x=0}^{1} \int_{y=-\sqrt{1-x^2}}^{\sqrt{1-x^2}} \int_{z=0}^{\sqrt{4-x^2-y^2}} x^2 \, dz \, dy \, dx$$

Do not evaluate the integral: just set it up carefully.

PROBLEM 4. Consider the region R in the plane defined by:

$$R_1^2 \leq x^2 + y^2 \leq R_2^2$$

A) Compute the moment of inertia of this unit-density plate rotated in the (x, y) plane about the origin.

B) Compute the radius of gyration of this (unit density) plate.

PROBLEM 5. Consider the region that is described in spherical coordinates as

$$0 \le \rho \le 2 \quad : \quad \frac{\pi}{2} \le \phi \le \pi \quad : \quad 0 \le \theta \le \frac{\pi}{2}$$

A) Describe carefully and/or draw a careful picture of this domain.

B) Compute the average of the function $f = 1/\rho$ on this domain.

PROBLEM 6. Consider the unit-density solid D (centered at the origin) given by the hemispherical region between a ball of radius 2 and a ball of radius 1, both with $z \ge 0$.

A) Compute $\bar{z}$, the z-coordinate of the centroid of D.

B) Does your answer to part (A) make sense?

PROBLEM 7. Consider the bounded region D in the plane given by

$$x^2 + y^2 \le R^2 \quad : \quad x \ge 0 \quad : \quad y \ge x$$

A) What is the area of this region?

B) Compute the centroid of D.

PROBLEM 8. Consider a joint probability density function on three variables, X, Y and Z, of the form $\rho(x,y,z) = e^{-Cz}(x^2 + y^2)$, where the domain D is the infinite cylinder where $x^2 + y^2 \le 1$ and $z \ge 0$.

A) What must the value of C be so that ρ is a probability density function on D?

B) What is the probability that $Z \le 1$?

PROBLEM 9. Consider the surface of a cone given by

$$z^2 = \frac{1}{3}\,(x^2 + y^2)\,.$$

A) Argue that this cone makes a 60° angle with the positive z-axis.

B) What is the volume element in spherical coordinates?

C) Compute the volume of the domain given by

$$x \ge 0\ ,\ y \ge 0\ ,\quad \frac{1}{4} \le x^2 + y^2 + z^2 \le 1\ ,\ z^2 \le \frac{1}{3}(x^2 + y^2)$$

PROBLEM 10. Compute the volume of the portion of the ball of radius 5 centered at the origin that satisfies $x^2 + y^2 \le 9$, using cylindrical coordinates.

PROBLEM 11. Use the change of variables

$$\begin{pmatrix} u \\ v \\ w \end{pmatrix} = F\begin{pmatrix} x \\ y \\ z \end{pmatrix} = \begin{pmatrix} x^2 + y^2 \\ y^2 - 2x^2 \\ 1 - \dfrac{z}{3} \end{pmatrix}$$

to convert to (u, v, w) coordinates the triple integral

$$\iiint_D x^3y + xy^3\, dx\, dy\, dz$$

Do not evaluate the transformed integral or worry about the limits of integration. Do convert the integrand and volume element in full detail.

PROBLEM 12. Consider the domain D satisfying

$$u \ge 0 \quad ; \quad v \ge 0 \quad ; \quad 0 \le u - v \le 2 \quad \text{and} \quad 4 \le u^2 + v^2 \le 9.$$

A) What would be an appropriate change of variables from (u, v) coordinates to (x, y) coordinates to make this domain *nice*?

B) Compute the integral

$$\iint_D u^2 - v^2 \, du \, dv$$

using the change of variables from part (A).

PROBLEM 13. Consider the following variation of cylindrical coordinates:

$$\begin{pmatrix} x \\ y \\ z \end{pmatrix} = F\begin{pmatrix} u \\ v \\ w \end{pmatrix} = \begin{pmatrix} 2u\cos v \\ 3u\sin v \\ w \end{pmatrix}$$

Use the Change of Variables Theorem to convert the volume element $dV = dx\,dy\,dz$ into (u, v, w) coordinates.

PROBLEM 14. Use the change of variables

$$\begin{pmatrix} u \\ v \\ w \end{pmatrix} = F\begin{pmatrix} x \\ y \\ z \end{pmatrix} = \begin{pmatrix} x^2 - y^2 \\ z(x+y) \\ 1 + 2z \end{pmatrix}$$

to convert to (u, v, w) coordinates the triple integral

$$\iiint_D z^2(x+y)^3(x-y) \, dx \, dy \, dz$$

Do not evaluate the transformed integral or give the limits of integration.

PROBLEM 15. Use the change of variables

$$\begin{pmatrix} u \\ v \\ w \end{pmatrix} = F\begin{pmatrix} x \\ y \\ z \end{pmatrix} = \begin{pmatrix} z - x \\ y^3 - x^3 \\ y + z \end{pmatrix}$$

to convert to (u, v, w) coordinates the triple integral

$$\iiint_D (x^2 + y^2)(x - z) \, dx \, dy \, dz$$

Do not evaluate the transformed integral or give the limits of integration.

PROBLEM 16. Consider the domain D given by

$$3 + x^3 \leq y \leq 5 + x^3 \quad \text{and} \quad 1 \leq xy \leq 2, \quad x, y > 0.$$

A) If we let $s = y - x^3$, what is a good choice for t so that so that the bounds on s and t for D are all constants?

B) Use the Change of Variables Theorem to express the area element $ds\,dt$ in terms of $dx\,dy$.

C) Use parts (A) and (B) to compute the following integral:

$$\iint_D \frac{3x^2}{y} + \frac{1}{x} \, dx \, dy$$

PROBLEM 17. Consider the domain D given by

$$\ln x \leq y \leq 1 + \ln x \quad \text{and} \quad 1 \leq xy^3 \leq 3, \quad x, y > 0.$$

A) If we let $u = y - \ln x$, what is a good choice for v so that so that the bounds for u and v for D are all constants?

B) Use the Change of Variables Theorem to express the area element $du\,dv$ in terms of $dx\,dy$.

C) Compute the integral of $e^y(3y^5 + y^6)$ on D.

PROBLEM 18. Compute the integral

$$\iint_D \frac{1}{x}\,dx\,dy$$

where D is the domain given by the inequalities $1 \leq y/x \leq 2$ and $1 \leq x + y \leq 4$.

PROBLEM 19. Compute the area of the domain D in the plane given by the inequalities $-1 \leq x^2 - y^2 \leq 1$ and $1 \leq x + y \leq 10$.

PROBLEM 20. Let D be the surface in 3-D parametrized as follows:

$$\begin{pmatrix} x \\ y \\ z \end{pmatrix} = S\begin{pmatrix} u \\ v \end{pmatrix} = \begin{pmatrix} 2+2u \\ u-v \\ 1-3v \end{pmatrix} \quad : \quad 0 \leq u \leq 3\ ;\ -1 \leq v \leq 1$$

A) What is the surface area element $d\sigma$ of this surface?

B) Use this to compute the surface area of D.

PROBLEM 21. Let S be the surface in 3-D parametrized as follows:

$$\begin{pmatrix} x \\ y \\ z \end{pmatrix} = F\begin{pmatrix} u \\ v \end{pmatrix} = \begin{pmatrix} 1-u+v \\ 2+3u-2v \\ 5+2u-v \end{pmatrix} \quad : \quad u^2 + v^2 \leq 1$$

A) Compute and simplify the surface area element $d\sigma$ of this surface.

B) Use this to compute the surface area of S.

PROBLEM 22. Consider the parametrized surface in 3-D given by

$$S\begin{pmatrix} u \\ v \end{pmatrix} = \begin{pmatrix} x \\ y \\ z \end{pmatrix} = \begin{pmatrix} v^2 \\ u^2 - v^2 \\ u^2 \end{pmatrix} \quad : \quad 0 \leq u \leq 2 \quad : \quad 0 \leq v \leq 3$$

A) Compute and simply the surface area element $d\sigma$, showing all steps.

B) Set up but do not solve a surface integral to compute the z-coordinate of the centroid of this surface.

ANSWERS & HINTS

PROBLEM 1. the volume is $8\pi/3$

PROBLEM 2. the volume is $2\pi/3$

PROBLEM 3.

$$\int_{x=0}^{1}\int_{y=-\sqrt{1-x^2}}^{\sqrt{1-x^2}}\int_{z=0}^{\sqrt{4-x^2-y^2}} x^2\,dz\,dy\,dx = \int_{\theta=-\frac{\pi}{2}}^{\frac{\pi}{2}}\int_{r=0}^{1}\int_{z=0}^{\sqrt{4-r^2}} r^4\cos^2\theta\,dz\,dr\,d\theta$$

PROBLEM 4. A) $I = \frac{1}{2}\pi(R_2^4 - R_1^4) = \frac{1}{2}M(R_2^2 + R_1^2)$; B) $R_g = \sqrt{\frac{(R_1^2 - R_2^2)}{2}}$

PROBLEM 5. this is one-eighth of a ball of radius two, with

$$\bar{f} = \frac{3}{4\pi}\int \frac{1}{\rho}\,dV = \frac{3}{4}$$

PROBLEM 6.

$$\bar{z} = \frac{1}{V}\int_D z\,dV = \left(\frac{3}{14\pi}\right)\left(\frac{15\pi}{4}\right) = \frac{45}{56}$$

PROBLEM 7. A) $A = \pi R^2/8$; B)

$$\bar{x} = \frac{4R}{3\pi}\left(2-\sqrt{2}\right) \ : \ \bar{y} = \frac{4R}{3\pi}\sqrt{2}$$

PROBLEM 8. A) $C = \pi/2$; B) $\mathbb{P} = 1 - e^{-\pi/2}$

PROBLEM 9. C)

$$V = \int_0^{\frac{\pi}{2}}\int_{\frac{\pi}{3}}^{\frac{\pi}{2}}\int_{\frac{1}{2}}^{1} \rho^2 \sin\phi\, d\rho\, d\phi\, d\theta = \frac{7\pi}{96}$$

PROBLEM 10.

$$V = \int_0^{2\pi}\int_0^{3}\int_{-\sqrt{25-r^2}}^{\sqrt{25-r^2}} r\, dz\, dr\, d\theta = \frac{244\pi}{3}$$

PROBLEM 11. using $du\,dv\,dw = 4xy\,dx\,dy\,dz$,

$$\iiint x^3y + xy^3\, dx\, dy\, dz = \iiint \frac{u}{4}\, du\, dv\, dw$$

PROBLEM 12. A) $x = u - v$ and $y = u^2 + v^2$; B) using $dx\,dy = 2|u+v|\,du\,dv$,

$$\iint_D u^2 - v^2\, du\, dv = \int_4^9\int_0^2 \frac{x}{2}\, dx\, dy = 5$$

PROBLEM 13. using the Change of Variables Theorem,

$$dx\,dy\,dz = \left|\det\begin{bmatrix} 2\cos v & -2u\sin v & 0 \\ 3\sin v & 3u\cos v & 0 \\ 0 & 0 & 1\end{bmatrix}\right| = 6|u|\,du\,dv\,dw$$

PROBLEM 14. using the Change of Variables Theorem,

$$\iiint_D z^2(x+y)^3(x-y)\, dx\, dy\, dz = \iiint \frac{uv}{4}\, du\, dv\, dw$$

PROBLEM 15. using the Change of Variables Theorem,

$$\iiint_D (x^2+y^2)(x-z)\, dx\, dy\, dz = \iiint -\frac{u}{3}\, du\, dv\, dw$$

PROBLEM 16. A) $t = xy$; B) $ds\,dt = (3x^3 + y)\,dx\,dy$; C)

$$\iint_D \frac{3x^2}{y} + \frac{1}{x}\, dx\, dy = \int_1^2\int_3^5 \frac{ds\,dt}{t} = 2\ln 2$$

PROBLEM 17. A) $v = xy^3$; B) $du\,dv = (3y^2 + y^3)dx\,dy$; C)

$$\iint_D e^y(3y^5 + y^6)\, dx\, dy = \int_1^3\int_0^1 e^u v\, du\, dv = 4(e-1)$$

PROBLEM 18. with $u = y/x$ and $v = x + y$, compute $du\,dv = |x+y|/x^2 dx\,dy$, yielding

$$\int_1^4\int_1^2 \frac{1}{1+u}\, du\, dv = 3\ln\frac{3}{2}$$

PROBLEM 19. with $u = x^2 - y^2$ and $v = x + y$, compute $du\,dv = 2|x+y|dx\,dy$, yielding

$$\int_1^{10}\int_{-1}^{1} \frac{u}{2v}\, du\, dv = \ln 10$$

PROBLEM 20. A) $d\sigma = 7\ du\ dv$; B) 42

PROBLEM 21. A) $d\sigma = \sqrt{3}\ du\ dv$; B) $\pi\sqrt{3}$

PROBLEM 22. A) $d\sigma = 4\sqrt{3}\ uv\ du\ dv$; B)

$$\bar{z} = \frac{\int_0^2 \int_0^3 4\sqrt{3}\ u^3 v\ dv\ du}{\int_0^2 \int_0^3 4\sqrt{3}\ uv\ dv\ du}$$

VOLUME IV : FIELDS

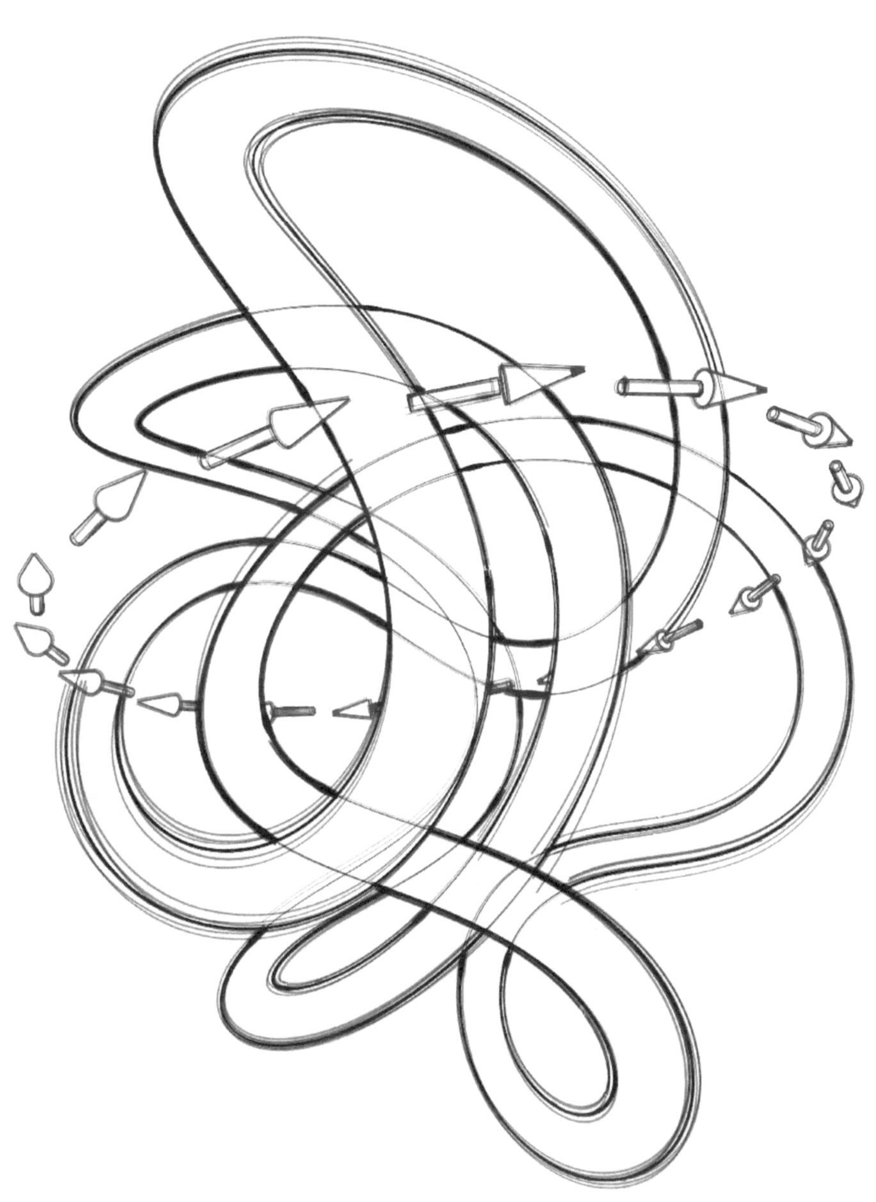

Week 12 : Path Integrals

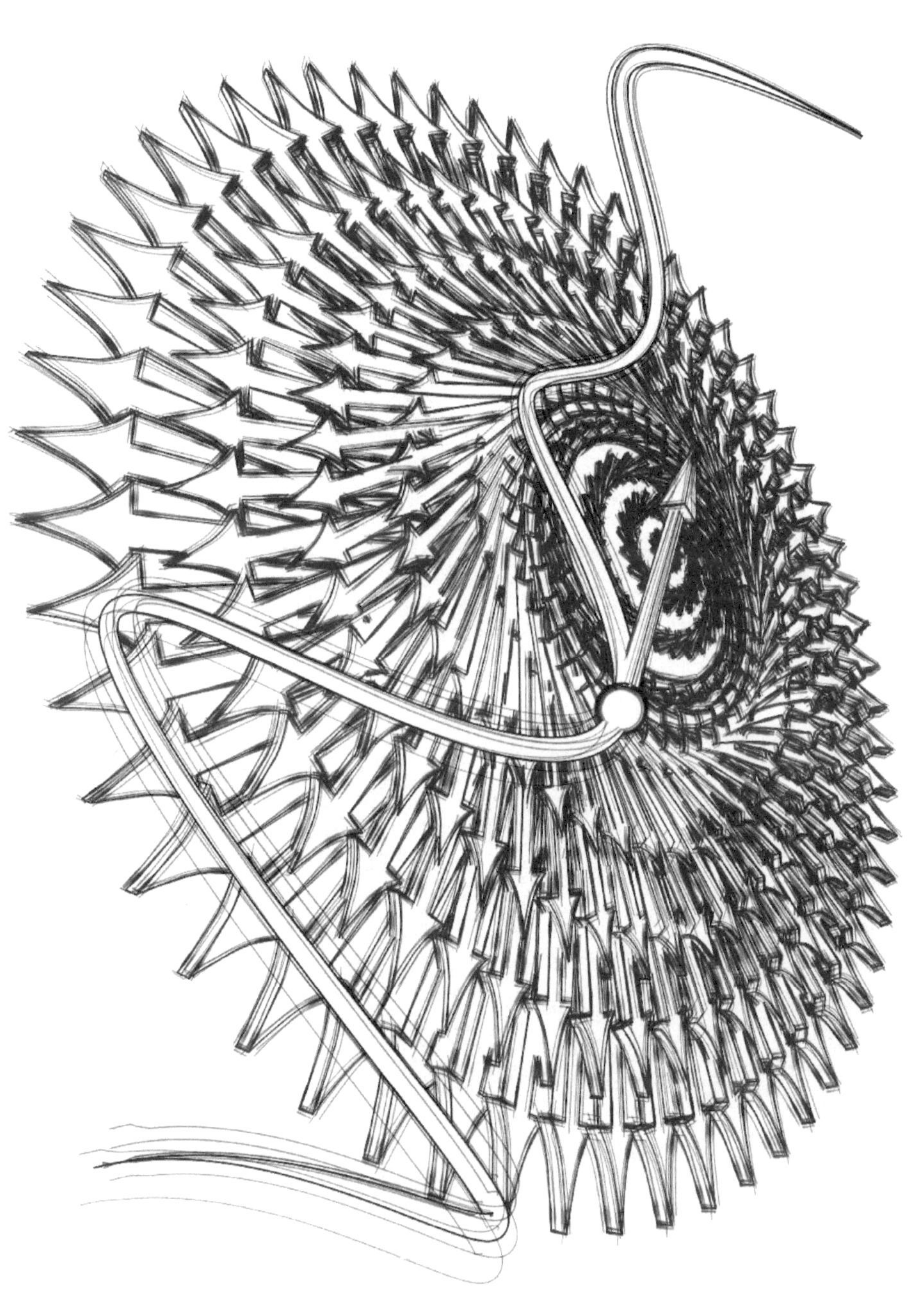

OUTLINE

MATERIALS: Calculus BLUE : Vol 4 : Chapters 1-5

TOPICS:

- ▷ Scalar and vector fields
- ▷ 1-forms and 1-form fields
- ▷ Scalar path integrals
- ▷ Gradient 1-form fields
- ▷ Scalar path integrals
- ▷ Path integrals and 1-form fields
- ▷ Independence of Path Theorem and potential functions
- ▷ Work and flux 1-forms and path integrals

LEARNING OBJECTIVES:

- ▷ Identify different types of fields : scalar, vector, 1-form
- ▷ Set up and compute scalar path integrals
- ▷ Evaluate 1-forms and 1-form fields on $\mathbb{R}^n$
- ▷ Integrate 1-form fields over parametrized paths
- ▷ Identify gradient 1-form fields
- ▷ Determine a potential function from a gradient 1-form field
- ▷ Use the Independence of Path Theorem to compute path integrals
- ▷ Interpret work and flux 1-forms in the plane

PRIMER

This begins the fourth quarter of our story, concerning *fields*.

FIELDS. A *scalar field* is nothing more than the usual scalar-valued functions we have been working with all along: $f: \mathbb{R}^n \to \mathbb{R}$. From now on, we are going to think of this as a *field* – an assignment of a scalar value to every point in the domain. There are other types of fields, determined by what type of object is assigned to every point in the domain. For example, the gradient of f, ∇f, is an example of a *vector field*: at each point in space a vector is assigned. Vector fields are very useful in Mathematics (geometry and differential equations in particular). They also hold a prominent role in Physics: electric, gravitational, and velocity fields are all important vector fields.

SCALAR PATH INTEGRALS. We begin the calculus of fields with integration. Given a scalar field f on $\mathbb{R}^n$, we already know how to integrate it with respect to the volume element dV over an n-dimensional domain $D \subset \mathbb{R}^n$. Consider what happens with a path $\gamma: [a, b] \to \mathbb{R}^n$. If we try to integrate f over the image of γ with respect to dV, we get zero (unless $n = 1$ of course). However, it should be possible to integrate the scalar field f over the path with respect to the "internal" arclength element. This can be done by pulling back the scalar field to the parameter domain of γ. We define the scalar path integral of f over γ as follows:

$$\int_\gamma f\, d\ell = \int_{t=a}^{b} f(\gamma(t))dt\,.$$

The crucial result - which is an immediate consequence of the Change of Variables Theorem of last week - is that this integral is *independent of the parametrization* of the path. Only the geometric path matters.

1-FORMS. Recall from Week 7 that for a scalar field f on $\mathbb{R}^n$, we can interpret the *gradient* ∇f as a vector field. One could likewise interpret the derivative $[Df]$ as a "matrix field," since the linear transformation depends on the point of evaluation. There is one more type of object associated with the derivative of a scalar field: recall also from Week 7 our use of *differentials* and our suspicious lack of formal definitions. We invoked "implicit differentiation" to make sense of

$$df = \frac{\partial f}{\partial x_1}dx_1 + \frac{\partial f}{\partial x_2}dx_2 + \cdots + \frac{\partial f}{\partial x_n}dx_n\,.$$

This is now to be classified as a new type of field: a *1-form field.* On $\mathbb{R}^n$, the *basis 1-forms* dx_i are objects that accept a vector in $\mathbb{R}^n$ and return the ith component: they are *projections* onto the x_i axis. One can combine these basis forms linearly, in the same way that one combines basis vectors. By having coordinate-dependent coefficients in front of the basis 1-forms, one obtains a 1-form field that varies from point-to-point. The gradient 1-form field df is such an object; other 1-form fields are not of the form df for any function f (see *Independence of Path*, below).

INTEGRATING 1-FORMS. Of the many ways to think about 1-form fields - as objects like vector fields or (locally) like gradients - there is one pseudo-definition that is tied to integration: a 1-form field α is an object that wishes to be integrated over a path. It is no coincidence that the integrands from single-variable calculus are 1-form fields on $\mathbb{R}^1$ of the form $f(x)dx$, and we know how to integrate such (via a limit of Riemann sums). For a general 1-form field α on $\mathbb{R}^n$ and a given parametrized path $\gamma\colon [a,b] \to \mathbb{R}^n$, one defines the integral via:

$$\int_\gamma \alpha = \int_{t=a}^{b} \alpha|_{\gamma(t)}(\gamma'(t))dt\,.$$

That is, at each point along the path, one feeds the velocity vector of the path to the 1-form evaluated at that point. Add these values up along the path, and this is the integral. This is independent of the parametrization, thanks to the Change of Variables Theorem. Changing the orientation (one could write $-\gamma$) is equivalent to reversing the parameter $t \mapsto -t$ and, since $dt \mapsto -dt$, this yields a minus sign.

INDEPENDENCE of PATH THEOREM. The first fundamental theorem of this Volume ties together 1-form fields, derivatives, and integrals. It states that for a gradient 1-form field df and a path $\gamma\colon [a,b] \to \mathbb{R}^n$, the path integral is

$$\int_\gamma df = f(\gamma(b)) - f(\gamma(a)).$$

This is, of course, the Fundamental Theorem of Integral Calculus, writ in terms of 1-form fields and path integrals.

When is a 1-form field a gradient? This is nontrivial. In single-variable calculus, every [integrable] function $f(x)$ is the derivative of the definite integral $\int_a^x f(t)dt$; thus, every 1-form field on $\mathbb{R}^1$ is a gradient 1-form. Beginning with dimension two, there are (many!) 1-form fields α which cannot be expressed as a gradient. The simple example $\alpha = y\,dx - x\,dy$ is illustrative of the general case: if $\alpha = df$, then $y = \partial f/\partial x$ and $-x = \partial f/\partial y$. However, this cannot be, since then the second partial derivatives do not match:

$$-1 = \frac{\partial}{\partial x}(-x) = \frac{\partial}{\partial x}\left(\frac{\partial f}{\partial y}\right) = \frac{\partial^2 f}{\partial x\,\partial y} \neq \frac{\partial^2 f}{\partial x\,\partial y} = \frac{\partial}{\partial y}\left(\frac{\partial f}{\partial x}\right) = \frac{\partial}{\partial y}(y) = 1$$

It is a fascinating result that this matching of partial derivatives is an if-and-only-if obstruction to being a gradient. If α is a 1-form field all pairwise partial derivatives match, then $\alpha = df$ is the gradient of some scalar field [a "*potential*"]. The problem of finding such a potential is an interesting challenge of computing the antiderivatives of the components of α and trying to match them up to a consistent scalar field. Is that approach better than taking an educated guess and checking whether it works? Any potential that works is a good potential.

WORK & FLUX. Why are we computing path integrals? There are two motivations for doing so in the plane, where we will initially focus. Given a planar vector field $\vec{F}$, consider the following two strangely (skew-)symmetric 1-form fields:

- ▷ *Work 1-form* : $\alpha_{\vec{F}} = F_x dx + F_y dy$.
- ▷ *Flux 1-form* : $\phi_{\vec{F}} = F_y dx - F_x dy$.

Integrating the work 1-form $\alpha_{\vec{F}}$ over a curve measures the work done by the vector field $\vec{F}$ along the curve: think of how the wind impacts the time and fuel needs of a cross-country flight. Integrating the flux 1-form $\phi_{\vec{F}}$ across the curve measures how much the vector field $\vec{F}$ "pushes" stuff across the curve (where there is an orientation - the signs on the flux 1-form are chosen so that if you integrate $\phi_{\vec{F}}$ along a closed loop (*e.g.*, a circle) you get the *outward* flux of the vector field across the curve, measuring what flows from inside to outside. Work and flux are primary motivations for integrating form fields. The work 1-form extends naturally to vector fields on $\mathbb{R}^n$ via an analogous work 1-form. The flux 1-form does not extend to a 1-form on $\mathbb{R}^3$, since a curve does not separate space into an *inside* and *outside* the way it does in the plane. The problem of how to efficiently compute work and flux, as well as how to generalize flux forms to 3-D, will be addressed next week.

DISCUSSION

QUESTION 1: [*speaking to a full classroom*] "Can you collectively give an example of a vector field by using your arms as vectors?"

Students will point in random directions. Ask if this is a continuous vector field, whatever that might mean. Follow up by asking students to work together to make a continuous vector field. They will likely all point at the speaker; or perhaps all in a consistent direction. This is a very good way to discuss continuity of vector fields without belaboring the definitions.

QUESTION 2: What is the simplest parametrization of a straight line between two points in $\mathbb{R}^3$?

After figuring this out, redo the problem for arbitrary dimensions: how much harder was that?

QUESTION 3: Integrate the scalar field $f = e^{-x^2-y^2}$ over the circle of radius R about the origin. What value of R maximizes this integral?

This leads to some interesting discussions: can you maximize without doing all the work explicitly? This foreshadows Green's theorem.

QUESTION 4: Can you think of an example of a scalar field whose integral would be finite on the hyperbola $xy = 1$ where $x, y \geq 0$?

This is good for reviewing asymptotic thinking and the arclength element. When students suggest using zero or mixed positive/negative values, refine the question to strictly positive fields.

QUESTION 5: Compute the centroid of the helical arc given by

$$\gamma(t) = \begin{pmatrix} \cos t \\ \sin t \\ t \end{pmatrix} \quad : \quad 0 \leq t \leq \pi$$

How much of this can be determined without doing the scalar path integral?

QUESTION 6: Compute the integral of $y^2 dx$ over the graph of $y = x^2$ as $-1 \leq x \leq 1$. What happens if instead you use the straight path between endpoints? Explain the difference between the answers. Which was easier to compute?

This is a good problem for emphasizing path dependence, as well as building intuition for what integration of a 1-form means: one is adding up values of y^2 times the infinitesimal change in x.

QUESTION 7: Integrate the 1-form field $x\,dy$ along a path in the plane from $(0,0)$ to (a, b) by (1) a straight line; and (2) a horizontal line followed by a vertical line.

This has a bit of foreshadowing for Green's Theorem. It is very good to dwell on the observation that the horizontal path is "invisible" to dy. The next question is a good follow-up.

QUESTION 8: Consider again the 1-form field $x\,dy$ and think of it as a "sensor" that detects someone moving along a path. Given any two points in the plane, is it possible to design a path between them so that $x\,dy$ never detects the motion?

With the previous problem done, the horizontal component is clear, and the vertical component along the y-axis is not hard to figure out either. Ask students as a follow-up if such motion is possible for a typical or even arbitrary 1-form field in the plane.

QUESTION 9: Compute the integral of $x^2dx + yz\,dy + \frac{1}{2}y^2dz$ over the path from the origin to $(0,0,10)$ given by $\gamma(t)$ where $x(t) = e^t \sin 4\pi t$; $y(t) = t(t-1)\cos^2 t$, and $z(t) = 10t^{10}$ for $t = 0 \ldots 1$.

Moral: whenever the integral looks impossible, look to a Theorem to get you out of trouble...

QUESTION 10: Integrate the 1-form field on $\mathbb{R}^n$ given by $\sum_i x_{i+1} dx_i$ (cyclic ordering) over the straight path from $\underline{0}$ to $\underline{1}$.

What does this integral measure? Does this integral exhibit path independence?

QUESTION 11: Is the following vector field a gradient?

$$\vec{V} = (e^{xy} + xye^{xy})\hat{\imath} + (x^2e^{xy} + ze^{-yz})\hat{\jmath} + (ye^{-yz})\hat{k}$$

What is its potential field? Use this to practice partial-integration versus outright guessing. In this case, it is probably easier to find the potential than it is to do the six partial derivatives to check for whether the potential exists.

QUESTION 12: Compute the flux of the vector field $\vec{F} = x\hat{\imath} + y\hat{\jmath}$ across the loop given by the square connecting (in order) the vertices $(0,0), (L, 0), (L, L), (0, L)$.

This is a good setup for Green's Theorem next week. This problem emphasizes the need to break this loop into segments, as well as the benefit in thinking before integrating, as the first and fourth path segments have the field tangent to the path & thus have no flux. One can also profitably compute the work along this loop as an exercise.

QUESTION 13: Compute the work W done by the vector field $\vec{F} = xy\,\hat{\imath} + yz\,\hat{\jmath} + xz\,\hat{k}$ along the straight-line path from $(1,2,0)$ to $(4,3,-1)$.

Can one make sense of computing the flux of the field along this curve? Why not? Why is it that work does make sense in 3-D or any dimension with a single (work) 1-form? Can one make sense of computing the flux of the field along this curve? Why not? Why is it that work makes sense in 3-D or any dimension by computing a single (work) 1-form?

QUESTION 14: How do you remember the formulae for work and flux 1-forms of a planar vector field $\vec{F} = F_x\hat{\imath} + F_y\hat{\jmath}$?

See if students can remember. The work is the easier of the two: $\alpha_{\vec{F}} = F_x\,dx + F_y\,dy$. For the flux 1-form $\phi_{\vec{F}} = F_x\,dy - F_y\,dx$, point out the lexicographic ordering of the x and y terms. Is this a trick? No, it is not, as shall be seen after doing differential forms in subsequent weeks.

ASSESSMENT PROBLEMS

PROBLEM 1. Compute the scalar path integral

$$\int_\gamma x^2\,d\ell$$

where γ is a path that follows the graph of the function $y = \ln x$ for $0 < x \leq \sqrt{3}$.

PROBLEM 2. Consider a unit-density semicircular wire γ given by $x^2 + y^2 = R^2$ with $y \geq 0$. The moments of inertia of this wire about the x- and y-axes are, respectively,

$$I_x = \int_\gamma y^2\,d\ell \quad : \quad I_y = \int_\gamma x^2\,d\ell$$

where $d\ell$ is the arclength element. Compute both these scalar path integrals and comment on which moment of inertia is larger, if there is a larger one.

PROBLEM 3. Consider the scalar field $f(x,y,z) = x^2 - y + z^2$.

A) Compute the scalar path integral

$$\int_\gamma f\,d\ell \quad : \quad \gamma(t) = \begin{pmatrix} \cos 2t \\ -2t \\ \sin 2t \end{pmatrix} \quad : \quad t = 0 \ldots \pi$$

B) Explain what happens to the value of this integral if the path is instead

$$\tilde{\gamma}(t) = \begin{pmatrix} \cos t \\ -t \\ \sin t \end{pmatrix} \quad : \quad t = 0 \ldots 2\pi$$

PROBLEM 4. Consider the scalar field in the plane $f(x,y) = x^2 + \frac{1}{2}y^2 - 1$, and the scalar path integral $I = \int_\gamma f\,d\ell$, where γ is the straight-line path from the origin to the point $(3,4)$.

A) Given an explicit parametrization of γ, using a parameter t.

B) Compute the value of I.

C) Is the value you found for I positive, negative, or zero; and does that make sense?

PROBLEM 5. Compute the scalar path integral

$$\int_{\gamma} x + y^2 + z^3 \, d\ell$$

over the straight-line path from the origin to the point $(6,3,-2)$.

PROBLEM 6. Consider the scalar field $f(x,y,z) = \sqrt{x} - y + z$.

A) Compute the scalar path integral

$$\int_{\gamma} f \, d\ell \quad : \quad \gamma(t) = \begin{pmatrix} t^2 \\ 1 - 2t^2 \\ 2t^2 \end{pmatrix} \quad : \quad t = 0 \ldots 1$$

B) Explain what happens to the value of this integral if you reverse the direction of the path; e.g., by letting t go from 1 to 0 instead.

PROBLEM 7. Consider the scalar field $f(x,y,z) = x^2 + y^2 + 3z$.

A) Compute the scalar path integral

$$\int_{\gamma} f \, d\ell \quad : \quad \gamma(t) = \begin{pmatrix} 2\cos t \\ -2\sin t \\ 3t \end{pmatrix} \quad : \quad t = 0 \ldots \pi$$

B) Explain what happens to the value of this integral if you reverse the direction of the path; e.g., by letting t go from π to 0 instead.

PROBLEM 8. Compute directly the work done by the vector field

$$\vec{F} = (x^2 z)\hat{\imath} + (yz^2)\hat{\jmath} + (xyz)\hat{k}$$

Along the path γ given by

$$\gamma(t) = \begin{pmatrix} x \\ y \\ z \end{pmatrix} = \begin{pmatrix} t^{1/2} \\ t^2 \\ t^{3/2} \end{pmatrix} \quad : \quad 0 \le t \le 2$$

PROBLEM 9. Use the Independence of Path Theorem to compute the work done by the gradient vector field

$$\nabla f = (yz(y-z))\hat{\imath} + (xz(2y-z))\hat{\jmath} + (xy(y-2z))\,\hat{k}$$

over the path γ given by

$$\gamma(t) = \begin{pmatrix} x(t) \\ y(t) \\ z(t) \end{pmatrix} = \begin{pmatrix} t^2 - 5t + 1 \\ 4t - 1 \\ t^2 - 2 \end{pmatrix} \quad : \quad 0 \le t \le 1$$

PROBLEM 10. Compute the integral

$$\int_{\gamma} y^2 dx + (2xy + y^2 + z)dy + y\,dz \quad : \quad \gamma(t) = \begin{pmatrix} t^2 \\ \sqrt{t} \\ 1 + t \end{pmatrix} \quad : \quad 0 \le t \le 4$$

PROBLEM 11. Consider the constant vector field $\vec{F} = 2\,\hat{\imath} - 3\,\hat{\jmath}$.

A) Write down the work 1-form $\alpha_{\vec{F}}$ and the flux 1-form $\phi_{\vec{F}}$ associated to $\vec{F}$.

B) Let γ be the typical counterclockwise unit circle in the plane. Compute both the flux of the field $\vec{F}$ across γ and the work done by $\vec{F}$ along γ.

C) Explain briefly why both the work and the flux were zero.

PROBLEM 12. Consider the vector field $\vec{F} = (y - 3x)\hat{\imath} + (x - 5y)\hat{\jmath}$.

A) What is the work 1-form $\alpha_{\vec{F}}$ associated to $\vec{F}$?

B) Let γ be a path that traces out a clockwise circle of radius 2 centered at the origin. Give an explicit parametrization of this path γ, using t as a parameter.

C) Compute the work done by the field $\vec{F}$ along the path γ.

PROBLEM 13. Consider the following 1-form field:

$$\alpha = y\,dx + xz\,dy + x^2\,dz$$

A) What is the value of α at $(2,3,-1)$ evaluated on the vector $v = \begin{pmatrix} -2 \\ 4 \\ 1 \end{pmatrix}$?

B) Integrate α over the path $\gamma(t) = \begin{pmatrix} t \\ t^2 \\ t^3 \end{pmatrix} : 0 \le t \le 1$.

C) This α is the work 1-form of what vector field on $\mathbb{R}^3$?

PROBLEM 14. Consider the vector field $\vec{F} = \left(x^2 - \frac{y^2}{x}\right)\hat{\imath} + (xy)\hat{\jmath}$.

A) Let γ be the path which goes from $(0,0)$ to $(1,1)$ along the graph of $y = x^k$, where $k > 0$ is a constant. Give an explicit parametrization of this path, using t as a parameter.

B) Compute the work done by the field $\vec{F}$ along the path γ (both as above).

C) What is the limit of this work as $k \to \infty$?

PROBLEM 15. Consider the following 1-form fields:

$$\alpha_1 = 2y\,dx - dy + x\,dz \quad : \quad \alpha_2 = z\,dx - x^2dy + y^2dz$$

A) Evaluate both these 1-form fields at the point $(-1,2,3)$.

B) Which of these 1-form fields has the smaller integral over the straight-line path from the origin to the point $(1,2,-2)$?

PROBLEM 16. Consider the 1-form field

$$\alpha = 3y\,dx + (3x + 2y)dy + 2z\,dz$$

A) What is the value of α at $(2,1,-1)$ evaluated on the vector $v = \begin{pmatrix} 1 \\ 2 \\ 0 \end{pmatrix}$?

B) Find a potential function f for α, so that $\alpha = df$.

C) Compute the integral of α along the straight-line path from (0,1,2) to (−1,2,3).

PROBLEM 17. Compute – directly or via finding a potential – the integral

$$\int_\gamma z\,dx - 2yz\,dy + (1 - y^2 + x)\,dz$$

over the straight-line path γ from $(0,1,2)$ to $(1,2,3)$.

PROBLEM 18. Consider the planar vector field $\vec{F} = y\,\hat{\imath} - x\,\hat{\jmath}$.

A) Draw a picture of this vector field and describe what it looks like.

B) What is the circulation of this vector field along the (counterclockwise) unit circle centered at the origin?

C) Explain why the flux of this vector field along the (counterclockwise) unit circle centered at the origin is exactly zero.

PROBLEM 19. Compute the integral

$$\int_\gamma ye^{xy}\,dx + (xe^{xy} - ze^{-yz})dy + (e^z - ye^{-yz})\,dz \quad : \quad \gamma(t) = \begin{pmatrix} t \\ 2t \\ 3t \end{pmatrix} \quad : \quad 0 \le t \le 1$$

PROBLEM 20. Compute the integral

$$\int_\gamma (z - y^2)dx + (2y - 2xy)dy + x\,dz \quad : \quad \gamma(s) = \begin{pmatrix} s^{2/3} \\ \sqrt{s/2} \\ s - 4 \end{pmatrix} \quad : \quad 0 \le s \le 8$$

PROBLEM 21. Consider the following planar vector field

$$\vec{F} = (C^2x - y)\hat{\imath} + (2Cx - Cy)\hat{\jmath}$$

which depends on some constant C. Which value of C produces zero work done by this field over the path given by the graph of $y = x^2$ for $-1 \le x \le 1$?

PROBLEM 22. Consider the following vector field on the (x, y) plane:

$$\vec{F} = (x^2 + 1)\hat{\imath} + (x + y)\hat{\jmath}$$

A) What is the value of $\vec{F}$ at the point $(3,1)$?

B) What is the work 1-form $\alpha_{\vec{F}}$ associated with $\vec{F}$?

C) What is the flux 1-form $\phi_{\vec{F}}$ associated with $\vec{F}$?

D) If you walk along a straight-line path from the origin to the point $(3,1)$, is the amount of work done by the field along the path positive, negative, zero, or undetermined?

PROBLEM 23. Consider the planar 1-form field α and vector field $\vec{V}$ given by

$$\alpha = 2y\,dx - x^2dy \quad : \quad \vec{V} = -x\,\hat{\imath} + y\,\hat{\jmath}$$

A) Evaluate both α and $\vec{V}$ at the point $(2,-1)$.

B) Is α the gradient 1-form of a potential function $f(x,y)$?

C) Draw a picture of $\vec{V}$ near the origin.

PROBLEM 24. Use the Independence of Path Theorem to integrate

$$\left(\frac{2x}{y} - 1\right)dx + \left(3y^2 - \frac{x^2}{y^2}\right)dy$$

over the path γ given by

$$\gamma(t) = \begin{pmatrix} x(t) \\ y(t) \end{pmatrix} = \begin{pmatrix} 1 + \arctan(t^2 - t) \\ t + 2\cos 3\pi t \end{pmatrix} \quad : \quad 0 \le t \le 1$$

PROBLEM 25. Consider the vector field $\vec{F} = (x^2 + y)\hat{\imath} + (2xy - 1)\hat{\jmath}$.

A) What is the work 1-form $\alpha_{\vec{F}}$ associated to $\vec{F}$?

B) Let γ be the path in the plane which goes from $(-1,0)$ to $(1,0)$ along the graph of $y = 1 - x^2$. Give an explicit parametrization of this path, using t as a parameter.

C) Compute the work done by the field $\vec{F}$ along the path γ.

PROBLEM 26. Use the Independence of Path Theorem to compute the work done by the vector field

$$\vec{F} = (1 - 2\sqrt{yz})\hat{\imath} + \left(2 - \frac{x\sqrt{z}}{\sqrt{y}}\right)\hat{\jmath} + \left(3 - \frac{x\sqrt{y}}{\sqrt{z}}\right)\hat{k}$$

over the path γ given by

$$\gamma(t) = \begin{pmatrix} x(t) \\ y(t) \\ z(t) \end{pmatrix} = \begin{pmatrix} (t-1)^2 \\ 1+3t \\ 4+5t \end{pmatrix} \quad : \quad 0 \le t \le 1$$

PROBLEM 27. Consider the vector field $\vec{F} = (x^2 + y^2)\hat{\imath} + (2xy - 1)\hat{\jmath}$.

A) What is the flux 1-form $\phi_{\vec{F}}$ associated to $\vec{F}$?

B) Let γ be the path in the plane which traces out a counterclockwise circle of radius 2 centered at the origin. Give an explicit parametrization of this path, using t as a parameter.

C) Compute the flux done by the field $\vec{F}$ across the path γ.

PROBLEM 28. Consider the vector field $\vec{F} = (x + y)\hat{\imath} + (x - y)\hat{\jmath}$.

A) What is the work 1-form $\alpha_{\vec{F}}$ associated to $\vec{F}$?

B) Let γ be the straight-line path in the plane which goes from $(-1,2)$ to $(3,0)$. Give an explicit parametrization of this path, using t as a parameter.

C) Compute the work done by the field $\vec{F}$ along the path γ.

ANSWERS & HINTS

PROBLEM 1. 7/3

PROBLEM 2. $I_x = I_y = \frac{1}{2}\pi R^3$

PROBLEM 3. A) $2\sqrt{2}\pi(1+\pi)$; B) same path, same integral

PROBLEM 4. A) $\gamma(t) = \begin{pmatrix} 3t \\ 4t \end{pmatrix}, 0 \le t \le 1$; B) $I = 70/3$

PROBLEM 5. $d\ell = 7\,dt$ and $\int_\gamma x + y^2 + z^3\,d\ell = 28$

PROBLEM 6. $d\ell = 6t\,dt$ and $\int_\gamma \sqrt{x} - y + z\,d\ell = 5$

PROBLEM 7. $d\ell = \sqrt{13}\,dt$ and $\int_\gamma x^2 + y^2 + 3z\,d\ell = \sqrt{13}\,\pi\left(4 + \frac{9}{2}\pi\right)$

PROBLEM 8. net work equals $\int_\gamma x^2z\,dx + yz^2dy + xyz\,dz = \frac{4}{3} + \frac{256}{7} + \frac{96}{11}\sqrt{2}$ (ugh)

PROBLEM 9. $36 - 6 = 30$, using the potential function $f = xyz(y - z)$

PROBLEM 10. $\int_\gamma y^2dx + (2xy + y^2 + z)dy + y\,dz = 230/3$

PROBLEM 11. work and flux both zero since constant field

PROBLEM 12. work equals zero: this is a gradient 1-form and the path is a loop

PROBLEM 13. A) -10 ; B) $19/15$; C) $y\hat{\imath} + xz\hat{\jmath} + x^2\hat{k}$

PROBLEM 14. work equals $\frac{1}{3} - \frac{1}{2k} + \frac{k}{2k+1}$ which limits to $\frac{5}{6}$ as $k \to \infty$

PROBLEM 15. A) $\alpha_1 = 4\,dx - dy - dz$; $\alpha_2 = 3\,dx - dy + 4\,dz$; B) both equal 1/2

PROBLEM 16. A) 19 ; B) $f = 3xy + y^2 + z^2$; C) 2

PROBLEM 17. -6, directly, or via potential $f = z(x - y^2 + 1)$

PROBLEM 18. A) ; B)

PROBLEM 19. $e^2 + e^{-6} + e^3 - 3$, via potential $f = e^{xy} + e^{-yz} + e^z$

PROBLEM 20. 4, via potential $f = z - xy^2 + y^2$

PROBLEM 21. $C = 1/4$ since integral evaluates to $\frac{1}{3}(8C - 2)$

PROBLEM 22. A) $\vec{F} = 10\hat{\imath} + 4\hat{\jmath}$; B) $\alpha_{\vec{F}} = (x^2 + 1)dx + (x + y)dy$; C) $\phi_{\vec{F}} = (x^2 + 1)dy - (x + y)dx$; D) positive

PROBLEM 23. A) $\alpha = -2\,dx - 4\,dy$; $\vec{V} = -2\hat{\imath} - \hat{\jmath}$; B) nope

PROBLEM 24. $-21/5$, via potential $f = x^2/y - x + y^3$

PROBLEM 25. the net work equals $\int_\gamma (x^2 + y)dx + (2xy - 1)dy = 14/15$

PROBLEM 26. 24, via potential $f = x - 2x\sqrt{yz} + 2y + 3z$

PROBLEM 27. the net flux equals $\int_\gamma (x^2 + y^2)dy - (2xy - 1)dx = 0$

PROBLEM 28. the net work equals $\int_\gamma (x + y)dx + (x - y)dy = 8$

Week 13 : Differential Forms

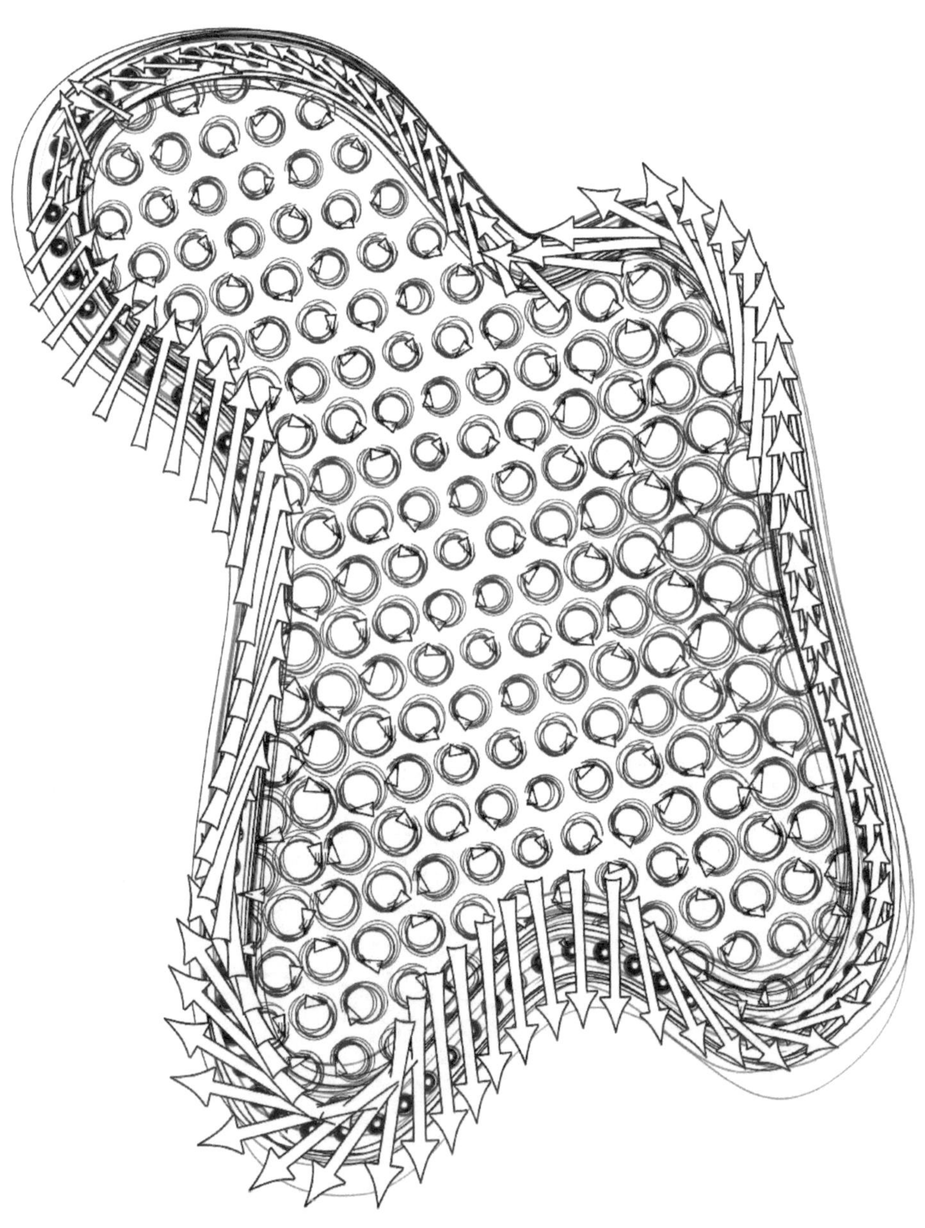

OUTLINE

MATERIALS: Calculus BLUE : Vol 4 : Chapters 6-8

TOPICS:

- ▷ Green's Theorem in the plane
- ▷ Work versus flux versions of Green's Theorem
- ▷ Path-dependence and orientation-dependence in Green's Theorem
- ▷ Curl and divergence of planar vector fields
- ▷ Curl and divergence of vector fields in 3-D
- ▷ Basis differential forms on $\mathbb{R}^3$ via determinants
- ▷ Differential form fields on $\mathbb{R}^3$
- ▷ Definition of the flux 2-form of a vector field in 3-D
- ▷ The wedge product $\wedge$ on forms and form fields
- ▷ The exterior derivative operator d on form fields
- ▷ Vanishing theorems for grad, curl, and div

LEARNING OBJECTIVES:

- ▷ Use Green's Theorem to compute 1-form path integrals
- ▷ Use Green's Theorem to compute work/flux of vector fields
- ▷ Orient boundaries of planar domains
- ▷ Interpret and compute curl and divergence of vector fields
- ▷ Evaluate basis k-forms on vectors via determinants
- ▷ Relate basis 2-forms to oriented projected areas
- ▷ Evaluate arbitrary form fields at points
- ▷ Compute and simplify the derivative d of a k-form field
- ▷ Compute and simplify the product $\wedge$ of a pair of form fields
- ▷ Use the vanishing theorems $\nabla \times \nabla f = 0$ and $\nabla \cdot \nabla \times \vec{F} = 0$

PRIMER

This is a critical week in which the first of our Fundamental Theorems – the Independence of Path Theorem – is generalized.

GREEN's THEOREM. Some of the more unusual coincidences that arise when computing integrals of 1-form fields over a loop. When integrating, e.g., $x\,dy$ over any simple closed loop in the plane, one seems to always obtain the area (up to sign, depending on the orientation of the curve). This points to a deep result.

Green's Theorem: If γ is the boundary of a domain $D \subset \mathbb{R}^2$, and f and g are C^1 (continuously differentiable) functions on D, then

$$\int_\gamma f\,dx + g\,dy = \iint_D \frac{\partial g}{\partial x} - \frac{\partial f}{\partial y}\,dA$$

Of course, in the case of $x\,dy$, the right-hand integrand evaluates to dA, yielding the area as double integral. It will be convenient to describe the relationship between the domain D and its boundary curve γ symbolically. For multiple reasons, we use the following cryptic shorthand: $\gamma = \partial D$, read "γ is the boundary of D." In the special case of a rectangular domain, the proof seems closely related to the Fundamental Theorem of Integral Calculus.

In the special case where we are computing circulation or flux of a planar vector field $\vec{F} = F_x\hat{\imath} + F_y\hat{\jmath}$, then Green's Theorem takes on two subtly symmetrical forms:

$$\text{circulation} = \int_\gamma F_x\,dx + F_y\,dy = \iint_D \frac{\partial F_y}{\partial x} - \frac{\partial F_x}{\partial y}\,dA$$

$$\text{flux} = \int_\gamma F_x\,dy - F_y\,dx = \iint_D \frac{\partial F_x}{\partial x} + \frac{\partial F_y}{\partial y}\,dA$$

The dual integrands on the right-hand sides are two very special types of derivatives associated to a planar vector field.

GRAD, CURL, DIV. Given a scalar field f on $\mathbb{R}^n$, we can understand its derivative in multiple ways: as the field of linear transformations $[Df]$; as the vector field ∇f; or as the 1-form field df. What does it mean to take the derivative of a vector field or a 1-form field? Let's begin with a planar vector field $\vec{F} = F_x\hat{\imath} + F_y\hat{\jmath}$. There are two distinct ways of differentiating such a field, each yielding a scalar field on $\mathbb{R}^2$.

$$\operatorname{curl}\vec{F} = \frac{\partial F_y}{\partial x} - \frac{\partial F_x}{\partial y} \quad : \quad \operatorname{div}\vec{F} = \frac{\partial F_x}{\partial x} + \frac{\partial F_y}{\partial y}$$

These have, thanks to Green's Theorem, interpretations in terms of infinitesimal circulation [curl] and infinitesimal flux [divergence]. Rotational or spinning vector fields have a nonzero curl (with sign denoting the orientation of the rotation). A positive divergence means that a vector field is locally expansive; local contraction is indicated by a negative divergence. With this new language, Green's Theorem is a local-to-global result: net circulation along a boundary is the sum of all the infinitesimal circulations (curl) over the interior; net flux across a boundary is the sum of all infinitesimal fluxes (divergence) over the interior.

Next week, we will lift Green's Theorem to 3-D. To do so, we need notions of curl and divergence for vector fields $\vec{F} = F_x\hat{\imath} + F_y\hat{\jmath} + F_z\hat{k}$. These are a bit more complex, as the curl is a vector field, but the divergence is a scalar field:

$$\operatorname{curl}\vec{F} = \nabla\times\vec{F} = \left(\frac{\partial F_z}{\partial y} - \frac{\partial F_y}{\partial z}\right)\hat{\imath} + \left(\frac{\partial F_x}{\partial z} - \frac{\partial F_z}{\partial x}\right)\hat{\jmath} + \left(\frac{\partial F_y}{\partial x} - \frac{\partial F_x}{\partial y}\right)\hat{k}$$

$$\operatorname{div}\vec{F} = \nabla\cdot\vec{F} = \frac{\partial F_x}{\partial x} + \frac{\partial F_y}{\partial y} + \frac{\partial F_z}{\partial z}$$

The interpretations of div and curl in 3-D are similar to the 2-D case. Divergence measures the expansion (positive) or contraction (negative) of volumes induced by the vector field. The curl has three components, each being a 2-D rotation in the three coordinate planes, as seen from the formula. Together, the curl vector gives an axis of infinitesimal rotation (the direction of the curl vector) and a strength of rotation (the length of the curl). The symbolic notation used is a holdover from Physics, where one imagines ∇ as a "*vector of partial differentiation operators*",

$$\nabla = \begin{pmatrix} \frac{\partial}{\partial x} \\ \frac{\partial}{\partial y} \\ \frac{\partial}{\partial z} \end{pmatrix},$$

with which one can dot or cross. It is good to know the notation, though a more modern approach exists and has several advantages.

DIFFERENTIAL FORMS. Why is it that we call 1-forms, *1-forms*? The prefix is a premonition of a deeper algebra of forms. For this and the next week, we will restrict attention to Euclidean $\mathbb{R}^3$ with a fixed basis of coordinates (x, y, z), sometimes working with (x_1, x_2, x_3) for generality.

The basis 1-forms are dx, dy, and dz (or dx_i in general): these take in a vector and return a scalar by projecting the vector to the x, y, or z axis respectively. Basis 1-forms generate (linear) 1-forms at a point and (nonlinear) 1-form fields on $\mathbb{R}^3$.

We define basis 2-forms on $\mathbb{R}^3$ using a *wedge product* notation: $dx \wedge dy$, $dy \wedge dz$, and $dz \wedge dx$. These objects eat an ordered pair of vectors and return in scalar in a manner best defined using determinants and indexed coordinates (x_1, x_2, x_3):

$$(dx_i \wedge dx_j)(\boldsymbol{u}, \boldsymbol{v}) = \det\begin{bmatrix} u_i & v_i \\ u_j & v_j \end{bmatrix}.$$

One thinks of the dx_i terms as meaning "return the x_i coordinate" and the wedge product $\wedge$ as "build a determinant". This definition yields a few simple algebraic rules for the wedge:

▷ Antisymmetry : $dx_i \wedge dx_j = -dx_j \wedge dx_i$ for all i, j ;
▷ Nilpotency : $dx_i \wedge dx_i = 0$ for all i.

Two-forms, being based on determinants, have geometric meaning: they capture *oriented projected area*. Given a pair of vectors $\boldsymbol{u}$ and $\boldsymbol{v}$, the value of $dx \wedge dy$ on the pair $(\boldsymbol{u}, \boldsymbol{v})$ is precisely the area of the parallelogram spanned by these vectors and projected to the (x, y) plane, with a +/- sign depending on orientation – which vector comes first.

Basis 2-forms in $\mathbb{R}^3$ lead to 2-form fields that vary from point-to-point. What would this be good for? Consider the problem of computing *flux* in 3-D. Given, say, a horizontal vector field $\vec{F} = F_x \hat{\imath}$, one can compute the flux through a small window of area; however, this is only detected by the projection of this window to the (y, z) plane. As well, since inside-vs-outside is also an orientation, flux in this case is a 2-form $F_x dy \wedge dz$. For a general vector field $\vec{F}$ in 3-D, the *flux 2-form*, $\Phi_{\vec{F}} = F_x dy \wedge dz + F_y dz \wedge dx + F_z dx \wedge dy$, captures the flow of a vector field $\vec{F}$ as a function of location, intensity, and the infinitesimal area patch defined by a pair of vectors $(\boldsymbol{u}, \boldsymbol{v})$ at a point. We will compute flux in the next Chapter.

Is that all? Almost. Following the pattern we have established, any k-form vanishes for $k > 3$, since one of the basis 1-forms must be repeated and the resulting determinant will have identical rows. However, one can make sense of a 0-form as being an object that eats zero vectors and returns a scalar – such an object is itself simply a scalar. A *0-form field* is a cognate of a scalar field.

ALGEBRA & CALCULUS of FORMS. Our goal is to do calculus with form fields. To do so, we must first master their algebra. Any two k-form fields on $\mathbb{R}^3$ can be added or subtracted. Multiplication is subtly oriented. Based on how we defined basis 2- and 3-forms via the $\wedge$ symbol, we define the wedge product $\alpha \wedge \beta$ of two form fields α and β via extension from basis forms. A j-form wedged with a k-form yields a $(j+k)$-form. One must be careful with signs, recalling that for 1-forms the wedge is antisymmetric: it is not necessarily so for other dimensions. Wedging with a 0-form field is the usual multiplication by a scalar field.

Differentiation and integration are the missing ingredients for a calculus of forms. The implicit differentiation operator d we have used throughout Calculus now ascends to a greater place.

There are a few interesting patterns which appear to tie together the gradient, curl, and divergence operators. One can check via direct computation that, for any scalar field f and any vector field $\vec{F}$ on $\mathbb{R}^3$,

$$\nabla \times \nabla f = 0 \quad \text{and} \quad \nabla \cdot (\nabla \times \vec{F}) = 0\,.$$

Is this a coincidence? Rewriting in terms of form fields deepens the mystery: for any 0-form field f and any 1-form field α,

$$d(df) = 0 \quad \text{and} \quad d(d\alpha) = 0\,.$$

This is often written in shorthand notation as $d^2 = 0$, where the superscript denotes composition of differentiation, and the "0" means the vanishing form field in the appropriate dimension. Such a simplification of complicated combinations of partial derivatives points to something deeper than coincidence.

DISCUSSION

QUESTION 1: What is the flux 1-form of $\vec{F} = (x^2+4y)\hat{\imath} + (x+y^2)\hat{\jmath}$? Use this to compute the flux of this vector field across the square in the plane with corners at $(-1,2)$ and $(2,5)$.

Try doing it without Green's Theorem and then with it... Beware of orientations! Which way is easier?

QUESTION 2: Compute the circulation of a fluid with velocity field $\vec{V} = (xy + y^2)\hat{\imath} + (x-y)\hat{\jmath}$ along the counterclockwise curve bounded by the graphs of $y = x^2$ and $x = y^2$.

This would require two path integrals, if you solved it with a path integral.

QUESTION 3: Integrate $\alpha = (\cos x + 3x^2y - 2y)dx + (x^3y + 4x - e^{2y})dy$ over a counterclockwise circle in the plane centered at (x_0, y_0).

When a problem seems unreasonably difficult (as this would be to compute directly), look for a Big Theorem to assist. The fact that the circle is arbitrarily positioned is a hint that the right hand side of Green's Theorem will have a simple – in this case constant – integrand.

QUESTION 4: Use Green's Theorem to show that the centroid of a region $D \subset \mathbb{R}^2$ is given by the path integrals

$$\bar{x} = \frac{1}{2A}\int_{\partial D} x^2\,dy \quad : \quad \bar{y} = \frac{1}{2A}\int_{\partial D} -y^2\,dx$$

where A is the area of D.

Students find this challenging, depending on how it is phrased. Write out the formulae for the centroid coordinates as double integrals; then use Green's Theorem to convert the putative right hand side path integrals. Then, look back at the original formulation. Why would anyone want to compute the centroid this way? Consider the problem of determining the centroid of a large complex parcel of land (say a dense forest) without satellite imagery. One could have a drone circumnavigate the boundary, keeping track of coordinates via GPS. Oh, but how can you get the area to normalize? Try to see if students can figure out how to use Green's Theorem to compute the area in a similar manner. This foreshadows Volume 4 Chapter 15 on applications to data science.

QUESTION 5: Draw a very complicated multiply-connected region in the plane and ask for the appropriate orientations at various points.

This leads to a lot of questions about how orientation works, as well as some misery.

QUESTION 6: (for students curious about the previous problem) Draw a very *very* convoluted simple closed curve in the plane. Pick a point somewhere deep inside the maze and ask "Is this on the inside or the outside?"

Hopefully, students will ask whether the question is well-defined. For students who can quickly navigate a maze in their head, ask what they would do if the maze-like curve were 10 or 100 times as large and complex. With a little prompting, students can figure out the algorithm of drawing a transverse curve to the outside and counting intersections mod 2. This is a good time to advertise other areas of Mathematics which curious students may wish to investigate.

QUESTION 7: Consider the planar vector field $\vec{F} = (ax + by)\hat{\imath} + (ay - bx)\hat{\jmath}$, for a and b constants. What are the curl and divergence of this field?

It is perhaps best to begin with the case where $b = 0$ or $a = 0$ and see what these special cases entail. This does not have an "ahha" solution – one explores until satisfied. As a follow-up question, what changes upon the addition of $+\, cz\, \hat{k}$ to the vector field (for c a constant)?

QUESTION 8: Consider the 2-form field $\beta = (x^2 - 2y)dy \wedge dz + (3y - 2z)dx \wedge dy$.

- ▷ Find a point (not at the origin) at which β vanishes. *(What does that mean?)*
- ▷ Find a point at which β is a positive multiple of $dx \wedge dy$.
- ▷ Find a point at which β is a multiple of $dx \wedge dz$. *(This is impossible: why?)*
- ▷ Find a point at which β equals $dy \wedge dz + dx \wedge dy$.
- ▷ Find a point at which $d\beta = dx \wedge dy \wedge dz$.

QUESTION 9: Consider the scalar fields $f = 3x + 2y - z$ and $g = 5y - 4z$. These have constant derivative 1-forms df, dg. What does the wedge $df \wedge dg$ mean?

This is very open-ended as worded. One can do a computation and move on; or, with more careful observation, one considers the wedge 2-form as the flux 2-form of some vector field. With a little hint, one sees the cross product emerge, and an answer to the old question from Week 2 about why the cross product only works in 3-D (but the wedge product works in all dimensions).

QUESTION 10: Compute the derivative of the 1-form

$$\alpha_{\vec{F}} = (x - y)dx + (y - z)dy + (z - x)dz$$

This 2-form field is, like every 2-form field, a flux 2-form field for some vector field $\vec{V}$: what is this $\vec{V}$?

It is, as per the lectures, the curl of $\vec{F}$; but it helps to do the computations directly.

QUESTION 11: Explain what is meant by the following claim: *every basis k-form is a determinant.*

It's best to start with $k = 3$, it which case the claim is clear. Then do $dx \wedge dy$: this eats a pair of vectors in $\mathbb{R}^3$, rips out the x and y components, then stacks them in a 2-by-2 matrix. How many ways are there to turn a pair of vectors into such a matrix?

End with the idea that the basis 1-forms are, trivially, determinants as well.

QUESTION 12: Follow-up to previous question: how do you interpret the wedge $\wedge$ in terms of determinants?

This is best begun in the context of going from 1-forms to 2-forms; then 2-forms to 3-forms. Try to get students to interpret the algebraic rules for $\wedge$ in terms of what we've learned about determinants in Week 4. In 3-D, it's a bit trivial: a clever student may ask whether you can wedge together 2-forms on $\mathbb{R}^4$ to get a nonzero 4-form.

QUESTION 13: Why is it that there are no nonzero 4-forms on $\mathbb{R}^3$?

Thinking in terms of determinants is a good idea, as ever.

QUESTION 14: How do you remember Green's Theorem? How is it related to what we have learned about 2-forms?

This is an essential problem to do live, since this is intentionally left out of the videotext. Get students to work through the derivative of the 1-form $\alpha = f\,dx + g\,dy$. What is the relationship between the 2-form $dx \wedge dy = -dy \wedge dx$ and the area form $dA = dx\,dy = dy\,dx$? Get to the point of being able to write down $\int_{\partial D} \alpha = \int_D d\alpha$ and hint that this is very close to the climax of the story. This is an excellent setup for Week 14.

ASSESSMENT PROBLEMS

PROBLEM 1. Consider the vector field $\vec{F} = (x^4 - 2y)\hat{\imath} + (y - y^3 + x^2)\hat{\jmath}$. Use Green's Theorem to compute the circulation of $\vec{F}$ along the circle given by the equation $x^2 + (y+3)^2 = 9$. Assume a counterclockwise orientation to the curve.

PROBLEM 2. Use Green's Theorem to compute the work done by the planar vector field $\vec{F} = (x^2 - 3y)\hat{\imath} + (2x + y^3)\hat{\jmath}$ along the path that:

1) starts at the origin;
2) follows the curve $y = x^2$ to $(1,1)$;
3) then follows the curve $y = \sqrt{x}$ back to the origin.

PROBLEM 3. State a version of Green's Theorem (any version is fine, as long as you explain what the various terms are).

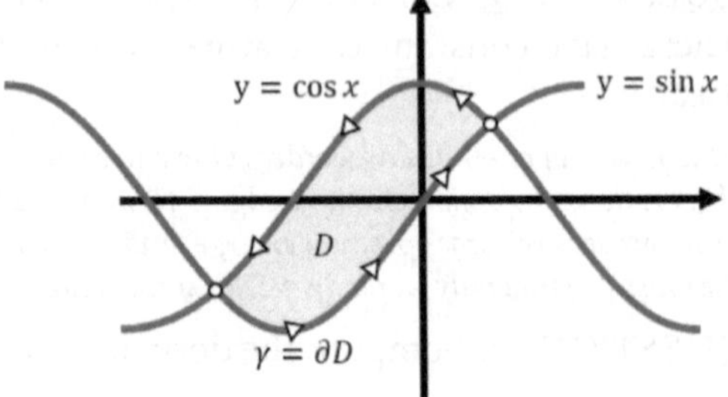

B) Use Green's Theorem to rewrite

$$\int_\gamma e^{xy}\,dx + \cos^2 3y\,dy$$

as an explicit double integral with careful bounds, for the curve γ as shown. Do not solve the double integral.

PROBLEM 4. Consider the vector field $\vec{F} = (x^3 - 3y)\hat{\imath} + (x^2 - y^3)\hat{\jmath}$. Use Green's Theorem to compute the circulation of $\vec{F}$ along the parametrized curve given by

$$\gamma(t) = \begin{pmatrix} x \\ y \end{pmatrix} = \begin{pmatrix} 3\cos(t^2) \\ 3\sin(t^2) \end{pmatrix} \quad : \quad 0 \le t \le \sqrt{2\pi}$$

PROBLEM 5. Use Green's theorem to compute the integral

$$\int_\gamma (x^2 + y^2 - 2y)dx + (2xy)dy$$

where γ is the (counterclockwise oriented) curve given by the boundary of the rectangle $0 \le x \le 2$, $-1 \le y \le 3$.

PROBLEM 6. Use Green's Theorem to compute the integral of the 1-form field

$$\alpha = \big(x\cos x - y(2 - 3x^2)\big)dx + (x^3 - 3y^4)dy$$

over the closed curve γ given by the boundary of the triangle with corners at $(-1,0)$, $(3,0)$, and $(3,5)$, following the points in that order.

PROBLEM 7. Use Green's Theorem to compute the integral of the 1-form field $\alpha = (\cos x + 2y)dx + (x - \sin y)dy$ over the curve γ parametrized as

$$\gamma(t) = \begin{pmatrix} x \\ y \end{pmatrix} = \begin{pmatrix} 1 + 3\cos(t) \\ 3\sin(t) - 2 \end{pmatrix} \quad : \quad 0 \le t \le 2\pi$$

PROBLEM 8. Use Green's Theorem to compute the circulation of the planar vector field $\vec{F} = (x^2y - x^5)\hat{\imath} + (y^3 - xy^2)\hat{\jmath}$ along the (counterclockwise) circle of radius two centered at the origin.

PROBLEM 9. Use Green's Theorem to compute the integral of the 1-form field $\alpha = (e^x - 2y)dx + (x^2 - y^2)dy$ over the closed curve γ given by the rectangle with corners at $(-1,-2)$, $(3,-2)$, $(3,1)$, and $(-1,1)$, following the points in that order.

PROBLEM 10. Compute the flux of the vector field

$$\vec{F} = (3xy - 5)\hat{\imath} + (x^2 + 4y)\hat{\jmath}$$

across the boundary of the domain in the plane satisfying $0 \le x \le 1 - y^2$. Assume a counterclockwise boundary.

PROBLEM 11. Consider the planar vector field

$$\vec{F} = x(1 - x^2 - y^2)\hat{\imath} + y(1 - x^2 - y^2)\hat{\jmath}$$

A) Let γ be the path which goes from $(R, 0)$ to $(-R, 0)$ along a semicircle of radius R in the upper half-plane where $y \ge 0$. Use Green's theorem to compute the flux of the field $\vec{F}$ across the path γ (using usual orientation).

B) For what value of $R > 0$ is this flux zero?

PROBLEM 12. Consider the vector field in 3-D

$$\vec{F} = (2x - y^2)\hat{\imath} + (z + x^2)\hat{\jmath} + (xy^2)\hat{k}$$

A) Compute and simplify the divergence $\nabla \cdot \vec{F}$ of this vector field.

B) What does the divergence you computed in part (A) tell you?

C) Compute and simplify the curl $\nabla \times \vec{F}$ of this vector field.

PROBLEM 13. Consider the vector field in 3-D

$$\vec{F} = (x - y)\hat{\imath} + (z + x)\hat{\jmath} + z\,\hat{k}$$

A) Compute the divergence $\nabla \cdot \vec{F}$ of this vector field.

B) Compute the curl $\nabla \times \vec{F}$ of this vector field.

C) What does your answer in (B) tell you about the field?

PROBLEM 14. Consider the following 1-form field

$$\alpha = (e^x + y^2)dx + (z - \sin x)dy - y\,dz$$

A) Compute and simplify the derivative $d\alpha$.

B) Compute the integral of α over the unit circle in the (y, z) plane.

PROBLEM 15. Let α and β be the following form fields on $\mathbb{R}^3$:

$$\alpha = xy\,dx - 2z^3dy + (x-y)^2dz \quad : \quad \beta = -3x\,dy \wedge dz + \frac{1}{z}dx \wedge dz$$

A) What is the value of β at the point $(2,5,-1)$?

B) What is the value of α at the point $(1,1,1)$ evaluated on the vector $v = \begin{pmatrix} 1 \\ 3 \\ -2 \end{pmatrix}$?

C) Compute and simplify $d\alpha$, the derivative of α.

D) Compute and simplify the product $\alpha \wedge \beta$.

PROBLEM 16. Let α and β be the following form fields on $\mathbb{R}^3$:

$$\alpha = 3\,dx + z\,dy + y^2dz \quad : \quad \beta = 2\,dy \wedge dz$$

A) What is the value of α at the point $(1, 2, 3)$?

B) Compute and simplify $d\alpha$, the derivative of α.

C) What does β measure?

D) Compute and simplify the product $\alpha \wedge \beta$.

PROBLEM 17. Consider the following fields on $\mathbb{R}^3$:

$$f = xy^2 - yz^2$$
$$\vec{V} = (x^2)\hat{\imath} + (xyz)\hat{\jmath} + (z^2 - 3xy)\hat{k}$$

A) Compute and simplify $\nabla \cdot \nabla f$, the divergence of the gradient of f.

B) Compute and simplify $\nabla(\nabla \cdot \vec{V})$, the gradient of the divergence of $\vec{V}$.

C) Compute and simplify $\nabla \times \nabla f$, the curl of the gradient of f.

PROBLEM 18. Consider the vector field

$$\vec{F} = (x+z)\hat{\imath} + (2y)\hat{\jmath} + (y-z)\hat{k}$$

A) Write out carefully the work 1-form $\alpha_{\vec{F}}$ associated with the field $\vec{F}$.

B) Write out carefully the flux 2-form $\Phi_{\vec{F}}$ associated with the field $\vec{F}$.

C) Compute the derivative $d(\alpha_{\vec{F}})$ of the work 1-form $\alpha_{\vec{F}}$ associated with the field $\vec{F}$.

D) What is the relationship between $\vec{F}$ and $d\alpha_{\vec{F}}$?

PROBLEM 19. Consider the following fields on $\mathbb{R}^3$:

$$g = xz - xyz^2$$
$$\vec{F} = (x^2y)\hat{\imath} + (y^2z)\hat{\jmath} + (x^3 + y^3 + z^3)\hat{k}$$

A) Compute and simplify $\nabla \cdot \nabla g$, the divergence of the gradient of g, if possible.

B) Compute and simplify $\nabla(\nabla \cdot \vec{F})$, the gradient of the divergence of $\vec{F}$, if possible.

C) Compute and simplify $\nabla \times \nabla g$, the curl of the gradient of g, if possible.

D) Compute and simplify $\nabla \times g$, the curl of g, if possible.

PROBLEM 20. Consider the following 1-form fields on $\mathbb{R}^3$:

$$\alpha = dx + (3x^2 - 4y)dy + e^z\, dz \quad : \quad \tilde{\alpha} = 2y\, dx + dy + x\, dz$$

A) Is there any point at which these two 1-forms are equal?

B) Compute and simplify the derivative $d\alpha$.

C) Compute and simplify the product $\tilde{\alpha} \wedge d\alpha$.

D) Why is $d\alpha \wedge d\tilde{\alpha} = 0$?

PROBLEM 21. Consider the following vector fields on $\mathbb{R}^3$:

$$\vec{F} = (x - y)\hat{\imath} + z\,\hat{\jmath} + (z - x)\hat{k} \quad : \quad \vec{G} = (x^2(y - z))\hat{\imath} - (xy^2 - z^2)\hat{\jmath} + (3xz^2)\hat{k}$$

A) Compute and simplify $\nabla \cdot \vec{G}$, the divergence of $\vec{G}$.

B) Compute and simplify $\nabla \times \vec{F}$, the curl of $\vec{F}$.

C) Compute and simplify $d\Phi_{\vec{G}}$, the derivative of the flux 2-form field of $\vec{G}$.

PROBLEM 22. Consider the following scalar fields on $\mathbb{R}^3$:

$$f = x^2y - y \quad : \quad g = 2xz + y^2$$

A) Compute the gradient 1-form fields df and dg.

B) Compute and simplify the product $df \wedge dg$ and show that it equals:

$$df \wedge dg = (4xy^2 - 2x^2z + 2z)\, dx \wedge dy + (2x^3 - 2x)\, dy \wedge dz - 4x^2y\, dz \wedge dx$$

C) Find a point not at the origin at which $df \wedge dg = 0$.

D) The product $df \wedge dg$ is the flux 2-form of some vector field $\vec{V}$. What is $\vec{V}$?

PROBLEM 23. Consider the 1-form fields

$$\alpha_1 = (y - e^{y^2})dx + x^3 \cos y\, dy \quad : \quad \alpha_2 = -3x^2 \sin y\, dx + 2xye^{y^2} dy$$

A) Compute the derivatives $d\alpha_1$ and $d\alpha_2$.

B) Integrate α_1 over the straight-line path from (0,0) to (1,0).

C) Let γ be the counterclockwise path around a circle of radius five centered at the origin. Which is bigger: $\int_\gamma \alpha_1$ or $\int_\gamma \alpha_2$?

PROBLEM 24. Let

$$f = xz^2 - y^2 \quad : \quad \alpha = 3\, dx - x^2 dy \quad : \quad \beta = x^2y\, dy \wedge dz + xy^2\, dz \wedge dx + dx \wedge dy$$

A) Fill in the blanks:

1) f is a ____-form
2) df is a ____-form
3) $d\beta$ is a ____-form
4) $df \wedge \alpha$ is a ____-form

B) Calculate and simply $d\beta$ as much as possible, showing work.

C) Calculate and simplify $df \wedge \alpha$ as much as possible, showing all steps below.

ANSWERS & HINTS

PROBLEM 1. by Green's Theorem

$$\int_{\partial D} (x^4 - 2y)dx + (y - y^3 + x^2)dy = \iint_D 2x + 2 \ dA = 18\pi$$

PROBLEM 2. by Green's Theorem

$$\int_{\partial D} (x^2 - 3y)dx + (2x + y^3)dy = \iint_D 5 \ dA = \frac{5}{3}$$

PROBLEM 3. by Green's Theorem

$$\int_{\partial D} e^{xy}dx + \cos^2 3y \, dy = \int_{-\frac{3\pi}{4}}^{\frac{\pi}{4}} \int_{\sin x}^{\cos x} -x \, e^{xy} \, dA$$

PROBLEM 4. by Green's Theorem

$$\int_{\partial D} (x^3 - 3y)dx + (x^2 - y^3)dy = \iint_D 2x + 3 \ dA = 27\pi$$

PROBLEM 5. by Green's Theorem

$$\int_{\partial D} (x^2 + y^3 - 2y)dx + (2xy)dy = \iint_D 2x + 2 - 3y^2 \ dA = -16$$

PROBLEM 6. by Green's Theorem

$$\int_{\partial D} \big(x \cos x - y(2 - 3x^2)\big)dx + (x^3 - 3y^4)dy = \iint_D 2 \ dA = 20$$

PROBLEM 7. by Green's Theorem

$$\int_{\partial D} (\cos x + 2y)dx + (x - \sin y)dy \ = \iint_D -1 \ dA = -9\pi$$

PROBLEM 8. by Green's Theorem

$$\int_{\partial D} (x^2y - x^5)dx + (y^3 - xy^2)dy \ = \iint_D -(x^2 + y^2) \ dA = -8\pi$$

PROBLEM 9. by Green's Theorem

$$\int_{\partial D} (e^x - 2y)dx + (x^2 - y^2)dy \ = \iint_D 2x + 2 \ dA = 48$$

PROBLEM 10. by Green's Theorem

$$\int_{\partial D} \phi_{\vec{F}} = \int_{\partial D} (3xy - 5)dy - (x^2 + 4y)dx \ = \iint_D 3y + 4 \ dA = 4A = \frac{64}{3}$$

PROBLEM 11. complete the arc to the half-disc, using a path along the x-axis from $(-R, 0)$ back to $(R, 0)$; there is zero flux along this line since $y = 0$; by Green's Theorem

$$\int_{\partial D} \phi_{\vec{F}} = \int_{\partial D} \big(y(1 - r^2)\big)dy - \big(x(1 - r^2)\big)dx \ = \iint_D 2 - 4r^2 \ dA = \pi R^2(1 - R^2)$$

which is zero along the unit-circular arc where $R = 1$.

PROBLEM 12. A) $\nabla \cdot \vec{F} = 2$; C) $\nabla \times \vec{F} = \ (2xy - 1)\hat{\imath} - y^2\hat{\jmath} + 2(x + y)\hat{k}$

PROBLEM 13. A) $\nabla \cdot \vec{F} = 2$; B) $\nabla \times \vec{F} = -\hat{\imath} + 2\hat{k} = \begin{pmatrix} -1 \\ 0 \\ 2 \end{pmatrix}$ constant

PROBLEM 14. A)

$$\begin{aligned} d\alpha &= (e^x \, dx + 2y \, dy) \wedge dx + (dz - \cos x \, dx) \wedge dy - dy \wedge dz \\ &= (-2y - \cos x)dx \wedge dy - 2dy \wedge dz \end{aligned}$$

B) by Green's Theorem in the (y,z) plane,

$$\int_{\partial D} (e^x + y^2)dx + (z - \sin x)dy - y\,dz = \iint_D -2\ dy\,dz = -2\pi$$

PROBLEM 15. A) $-6\,dy \wedge dz - dx \wedge dz$; B) -5 ; C/D)

$$d\alpha = -x\,dx \wedge dy + \left(6z^2 - 2(x-y)\right)dy \wedge dz + 2(x-y)dx \wedge dz$$

$$\alpha \wedge \beta = (2z^2 - 3x^2y)dx \wedge dy \wedge dz$$

PROBLEM 16. A) $3\,dx + 3\,dy + 4\,dz$; B) $d\alpha = (2y-1)dy \wedge dz$; C) twice the oriented projected area in the (y,z) plane ; D) $\alpha \wedge \beta = 6\,dx \wedge dy \wedge dz$

PROBLEM 17.

$$\nabla \cdot \nabla f = \nabla \cdot \left(y^2\hat{\imath} + (2xy - z^2)\hat{\jmath} - 2yz\hat{k}\right) = 2(x-y)$$

$$\nabla \cdot \vec{V} = 2x + xz + 2z$$

$$\nabla\left(\nabla \cdot \vec{V}\right) = (2+z)\hat{\imath} + (2+x)\hat{k}$$

$$\nabla \times \nabla f = 0$$

PROBLEM 18.

$$\alpha_{\vec{F}} = (x+z)dx + (2y)dy + (y-z)dz$$

$$\Phi_{\vec{F}} = (x+z)dy \wedge dz + (2y)dz \wedge dx + (y-z)dx \wedge dy$$

$$d\alpha_{\vec{F}} = dy \wedge dz - dx \wedge dz = \Phi_{\nabla \times \vec{F}}$$

PROBLEM 19.

$$\nabla \cdot \nabla g = \nabla \cdot \left((z - yz^2)\hat{\imath} - xz^2\hat{\jmath} + (x - 2xyz)\hat{k}\right) = -2yz$$

$$\nabla\left(\nabla \cdot \vec{F}\right) = \nabla(2xy + 2yz + 3z^2) = (2y)\hat{\imath} + (2x+2z)\hat{\jmath} + (2y+6z)\hat{k}$$

$$\nabla \times \nabla g = 0$$

PROBLEM 20. A) $\alpha = \tilde{\alpha}$ at $\left(1, \frac{1}{2}, 0\right)$; B) $d\alpha = 6x\,dx \wedge dy$; C) $\tilde{\alpha} \wedge d\alpha = 6x^2 dx \wedge dy \wedge dz$; D) no nonzero 4-forms on $\mathbb{R}^3$

PROBLEM 21.

$$\nabla \cdot \vec{G} = 2x(y-z) - 2xy + 6xz$$

$$\nabla \times \vec{F} = -\hat{\imath} + \hat{\jmath} + \hat{k} = \begin{pmatrix} -1 \\ 1 \\ 1 \end{pmatrix}$$

$$d\Phi_{\vec{G}} = (\nabla \cdot \vec{G})dx \wedge dy \wedge dz = (2x(y-z) - 2xy + 6xz)dx \wedge dy \wedge dz$$

PROBLEM 22. A) $df = 2xy\,dx + (x^2 - 1)dy$; $dg = 2z\,dx + 2y\,dy + 2x\,dz$; C) anything of the form $(0, C, 0)$ or $(1, 0, C)$; D) $\vec{V} = (2x^3 - 2x)\hat{\imath} - 4x^2y\hat{\jmath} + (4xy^2 - 2x^2z + 2z)\hat{k}$

PROBLEM 23. A)

$$d\alpha_1 = \left(3x^2\cos y + 2ye^{y^2} - 1\right)dx \wedge dy \quad : \quad d\alpha_2 = \left(3x^2\cos y + 2ye^{y^2}\right)dx \wedge dy$$

B) $\int \alpha_1 = -1$; C) by Green's Theorem and linearity

$$\int_{\gamma = \partial D} \alpha_2 - \alpha_1 = \iint_D d\alpha_2 - d\alpha_1 = \iint_D dx\,dy = 25\pi > 0$$

thus the integral of α_2 is larger

PROBLEM 24. A) 0, 1, 3, 2;

$$d\beta = 4xy\,dx \wedge dy \wedge dz$$

$$df \wedge \alpha = (-x^2z^2 + 6y)dx \wedge dy + 6xz\,dz \wedge dx + 2x^3z\,dy \wedge dz$$

Week 14 :
The Fundamental Theorem

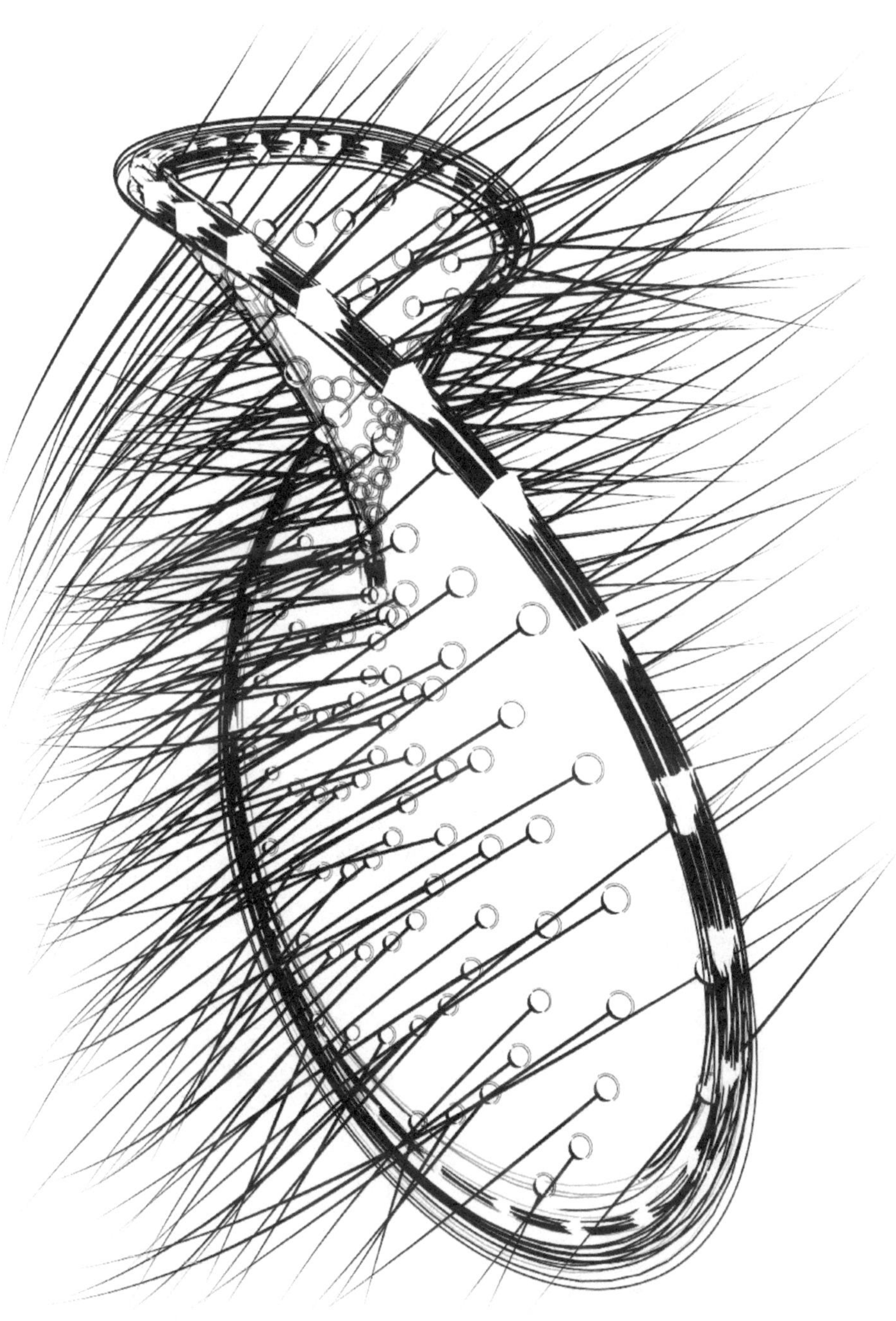

OUTLINE

MATERIALS: Calculus BLUE : Vol 4 : Chapters 9-12

TOPICS:

- ▷ Integration of 2-form fields over parametrized surfaces
- ▷ Flux of a vector field across a surface
- ▷ Gauss' Theorem for flux across a boundary surface
- ▷ Stokes' Theorem for circulation along a boundary loop
- ▷ Orientation and induced orientation on a boundary
- ▷ The differential forms version of Green/Gauss/Stokes/IoP
- ▷ The relationship between the Fundamental Theorems of Calculus

LEARNING OBJECTIVES:

- ▷ Integrate a 2-form field over a parametrized surface
- ▷ Interpret the integral of a 2-form as flux across an oriented surface
- ▷ Use Gauss' Theorem to simplify/compute integrals
- ▷ Use Stokes' Theorem to simplify/compute integrals
- ▷ Determine the induced boundary orientation of an oriented surface
- ▷ Choose the correct Fundamental Theorem to solve a given integral

PRIMER

This is the end of our story. The main results of this final week begin as generalizations of Green's Theorem from 2-D to 3-D. In the same way that there are two interpretations of Green's Theorem – work and flux – there are two distinct generalizations to 3-D. These entwine the parallel notions of curl and divergence, while tying together all we have learned about differential forms.

INTEGRATING 2-FORM FIELDS. In Week 12, we interpreted 1-form fields as measuring the work done by a vector field $\vec{F}$ in a particular direction. Given an oriented path γ, the 1-form takes in the tangent vector along the curve and returns a scalar – the work done by $\vec{F}$ along the tangent vector to the path. Thus, 1-form fields are integrated over paths and return net work.

Working with 2-form fields is similar. In 3-D, we interpret a 2-form field as the flux 2-form $\Phi_{\vec{F}}$ of a vector field $\vec{F}$. We do not compute flux across a curve in 3-D; rather, flux is computed across an oriented surface S. In the vector notation of Physics, one computes the flux of $\vec{F}$ across S as a surface integral (*cf.* Week 11) using the surface are element $d\sigma$ and a field $\hat{n}$ of unit vectors *normal* (this is, orthogonal) to the surface. The dot product between $\vec{F}$ and $\hat{n}$ gives infinitesimal flux, which can then be integrated over the surface. This will be our motivation to integrate 2-form fields:

$$\text{flux} = \iint_S \vec{F} \cdot \hat{n}\, d\sigma = \int_S \Phi_{\vec{F}}$$

In practice, is can be difficult to compute both $\hat{n}$ and $d\sigma$, and the direct integration of the flux 2-form field is often cleaner. For a parametrized surface given by $S: \mathbb{R}^2 \to \mathbb{R}^3$ with parameters s and t, the tangent plane to the surface is spanned by the two columns of $[DS]$ – the partial derivatives $\partial S/\partial s$ and $\partial S/\partial t$.

These are the vectors which (in that order) are taken in by the flux 2-form $\Phi_{\vec{F}}$, returning an infinitesimal flux across the surface at that point. By integrating this over the parameter plane (s, t) we obtain the net flux. The Change of Variables Theorem implies that only the surface matters, not the way in which it is parametrized (so long as orientations match).

There is no need to cast things exclusively in terms of flux. In the same way that work serves as motivation to define the integral of a 1-form field α over a path γ, one can define the integral of *any* 2-form field β over a surface parametrized by $S = S(s, t)$ by feeding the columns of the derivative $[DS]$ into β and integrating with respect to the area element $dA = ds\, dt$ in the parameter plane:

$$\int_S \beta = \iint \beta_S[DS]\, dA = \iint_{s,t} \beta_{S(s,t)}\left(\frac{\partial S}{\partial s}, \frac{\partial S}{\partial t}\right) ds\, dt$$

This looks more complex than it is – one simply integrates the values of β over the parameter plane. A single integral sign is used to denote the integral of a k-form field, as opposed to the single, double, or triple integral used on the parameter space when evaluating the integral explicitly.

THE GAUSS THEOREM. The flux form of Green's Theorem can be lifted to 3-D. The Physics/vector notation version of the theorem states that for a vector field $\vec{F}$ on $\mathbb{R}^3$, the flux of $\vec{F}$ across the (oriented) boundary of a solid domain D is the net divergence of $\vec{F}$ within the interior of D:

$$\text{flux} = \iint_{\partial D} \vec{F} \cdot \hat{n}\, d\sigma = \iiint_D \nabla \cdot \vec{F}\, dV = \text{net div}.$$

This replacement of a surface integral with a triple integral is often advantageous, given the complexities we have seen with surface area elements. One tradeoff comes in keeping track of orientations. For a solid domain D, the boundary surface ∂D has an *outward-pointing* unit normal vector field $\hat{n}$. This can be an issue with more complicated solid bodies having tunnels or interior cavities: the normal vector should always point from the inside to the outside.

When written in the language of differential forms, Gauss' Theorem becomes much easier to write and remember. Recall from last week that the derivative of a flux 2-form is the divergence of the vector field time the usual volume 3-form. The left-hand side of the Gauss Theorem becomes the integral of a flux 2-form field, and the right-hand side becomes the integral of its derivative:

$$\int_{\partial D} \Phi_{\vec{F}} = \int_D d\Phi_{\vec{F}}.$$

It is not a coincidence that this is the same as the differential forms version of Green's Theorem but in 3-D with 2-forms instead of 1-forms.

THE STOKES THEOREM. The circulation form of Green's Theorem can be lifted to 3-D. In Physics/vector notation, the theorem is stated thusly:

$$\text{circ} = \int_{\partial D} \vec{F} \cdot d\boldsymbol{x} = \iint_D (\nabla \times \vec{F}) \cdot \hat{n}\, d\sigma = \text{flux of curl}$$

This has a curious corollary. If you are tasked with computing the flux of the curl of a vector field (see next week's Epilogue for why this might happen in fluid dynamics or electromagnetics) across a complicated surface, you can simplify the surface as you wish, so long as the boundary remains the same. This is similar in spirit to the Independence of Path Theorem – integrating a derivative over the interior is the same as evaluating the principal on the boundary, so that only the boundary matters. In the case of Stokes' Theorem, we might call this feature an "*Independence of Surface*" result.

In the language of differential forms, we once again have a vast simplification:

$$\int_{\partial D} \alpha_{\vec{F}} = \int_{D} d\alpha_{\vec{F}} .$$

Unlike Gauss' Theorem (which usually goes in one direction – replacing a flux integral with a net divergence integral), Stokes' Theorem can be useful in either direction, to compute a circulation via a surface integral, or to compute a flux of a curl as a circulation. One must, as always, worry about the correct orientation.

THE FUNDAMENTAL THEOREM. Written in the language of differential forms, all the major theorems of vector calculus have the same form:

$$\int_{\partial D} \omega = \int_{D} d\omega .$$

This, the Generalized Stokes' Theorem, holds for any k-form field ω defined on an oriented $(k+1)$-dimensional domain D with oriented boundary ∂D. This is the end of our story and the goal to which we have worked. With the proper language of differential forms, the unity of Green, Gauss, Stokes, and the Independence of Path Theorems is manifest. They are all – in substance and in proof – the Fundamental Theorem of Integral Calculus:

$$\int_{[a,b]} df = \int_{\partial[a,b]} f = f(b) - f(a).$$

By interpreting the integral of a 0-form field f over a 0-dimensional point x as the evaluation $f(x)$; and by using $+/-$ orientation as a sign on the integral (as always done), the integral of f over the boundary of an interval $[a,b]$ is the evaluation at the "positive" endpoint b minus the evaluation at the negative endpoint a. This, the FTIC, is both the prototype of the generalized Stokes' Theorem and the core ingredient of its proof.

DISCUSSION

QUESTION 1: Compute the integral of the 2-form $\beta = z\, dx \wedge dy \ - \ x^2\, dy \wedge dz$ over the surface given by $z = 4 - x^2 - y^2$ with $z \geq 0$. Use the positive z-axis to orient.

Begin with parametrizing the surface? That is one approach and makes for a good exercise in the definitions; but consider the symmetry of this object and what the $dx \wedge dy$ versus $dy \wedge dz$ terms do on this particular surface. This can be integrated very nicely without explicit parametrization or worrying about integrating 2-forms at all. Why is the final answer related to an enclosed volume?

QUESTION 2: What is the integral of the 2-form field $\beta = e^{-z}\, dx \wedge dy$ over the paraboloid given by $z = x^2 + y^2$? Use the positive z-axis to orient.

This seems impossible, as the paraboloid extends infinitely; however, this gives a finite integral that can be easily computed either via polar coordinates or via remembering basic Gaussians.

QUESTION 3: What is the flux of $\vec{F} = x\hat{\imath} + y^2\hat{\jmath} + (z+y)\hat{k}$ across the boundary of the cylindrical solid within $x^2 + y^2 \leq 4$, below $z = 8$, and above $z = x$? Use an outward-pointing normal.

This includes all three boundary components and is clearly set up to be a Gauss Theorem problem. A little bit of reasoning with symmetry suffices to make short work of the triple integral.

QUESTION 4: What is the flux of the field $\vec{F} = y^2\hat{\imath} - x^2\hat{\jmath} + 2\hat{k}$ across the upper hemisphere of radius R at the origin? Orient it with the positive z-axis.

This is one of the few flux problems in which computing the unit normal field and taking the dot product is perhaps helpful, though this problem can be done in multiple ways. Try doing it with the direct computation (using the surface are element from spherical coordinates if needed); then, try with Gauss' Theorem by completing to a solid hemisphere, noting that the divergence is zero, then computing the flux across the disc in the (x, y) plane.

Since the divergence of $\vec{F}$ is zero, we can conclude that it is a curl: $\vec{F} = \nabla \times \vec{V}$ for some $\vec{V}$. Can you figure out what that field would be? (This is perhaps not so easy...) If so, one could get the flux of $\vec{F}$ by computing the circulation of $\vec{V}$ along the circle of radius R in the (x, y) plane.

QUESTION 5: Compute the following circulation:

$$\int_\gamma (e^z + 3y\cos x)dx + (3\sin x)dy + (2 + xe^z)dz$$

where the curve γ is parametrized as

$$\gamma = \begin{pmatrix} (5+\cos 3t\,)\cos t \\ (5+\cos 3t)\sin t \\ \sin t \end{pmatrix} \;;\; 0 \leq t \leq 2\pi$$

The first step is to notice that a direct computation of this circulation is practically impossible. This suggests trying Stokes' Theorem. However, that too seems daunting, as the curve is not easily seen to be the boundary of a "nice" surface. Perhaps some trick involving the independence of surface property of Stokes? No. The lesson here is to go ahead and compute $d\alpha$ (or the curl of the associated vector field) and see that it vanishes, obviating the need for dealing with the geometry of the surface.

QUESTION 6: Compute the circulation of the field $e^{-x}\hat{\imath} + e^x\hat{\jmath} + e^z\hat{k}$ along the triangle in the first octant (where $x, y, z \geq 0$) cut out by the plane $2x + y + 2z = 2$, oriented by traversing the vertices as they go from x-axis to y-axis to z-axis, then back to the x-axis.

Clearly a problem meant to suggest Stokes' Theorem, though it would be possible to do this directly, by parametrizing three straight-line segments and substituting. However, Stokes' is the superior play, as the curl of this vector field is $e^x\hat{k}$ (or, perhaps better, the derivative of the work 1-form equals $e^x dx \wedge dy$). In either case, it is clear that one must compute the flux across this triangular surface. Try to reason out why it suffices to integrate over the projection of this triangle to the (x, y) plane. Unlike many Green/Gauss/Stokes problems, the resulting integral is not immediately trivial to compute – the e^x along with the triangular domain is a good review of double integrals.

QUESTION 7: What is the flux of the field

$$\vec{F} = (\cos z + xy^2)\hat{\imath} + (x\,e^{-z})\hat{\jmath} + (\sin y + x^2 z)\hat{k}$$

out of the surface given by the paraboloid $z = x^2 + y^2$ satisfying $z \leq 4$.

This is a difficult problem since it is not set up directly for Gauss. Try computing the flux across the top disc at $z = 4$ directly (not trivial, but do use symmetry to eliminate the odd $\sin y$ term);

then use Gauss to get the total flux across both boundary components. As the divergence is $x^2 + y^2$, this problem uses cylindrical coordinates in an essential way and gives a good review.

QUESTION 8: Compute the flux of the curl of

$$\vec{F} = (y + x\sin x^2)\hat{\imath} + \left(x^2 + e^{y^2-5y}\right)\hat{\jmath} + (x^2 + y^2)\hat{k}$$

across the graph of the function $z = \cos^3(\pi(x^2 + y^2)/2)$ for $x^2 + y^2 \leq 1$, oriented by the positive z-axis.

The phrase "flux of the curl" is a trigger for Stokes' Theorem to be used. However, in this case, computing the circulation of $\vec{F}$ along the boundary is not going to work (as students should discover the hard way, time permitting). Computing the derivative of the work 1-form of $\vec{F}$ gives

$$d\alpha_{\vec{F}} = (2x - 1)dx \wedge dy + 2x\, dx \wedge dz + 2y\, dy \wedge dz\,.$$

This is helpful. Using the independence-of-surface property, one can integrate over the unit disc in the (x, y) plane. Symmetry and area give a quick clean answer.

QUESTION 9: Recall Week 13 Question 4, in which Green's Theorem was used to compute centroids of a 2-D domain based on information along the boundary:

$$\bar{x} = \frac{1}{2A}\int_{\partial D} x^2\, dy \quad : \quad \bar{y} = \frac{1}{2A}\int_{\partial D} -y^2\, dx \quad : \quad A = \int_{\partial D} x\, dy = -\int_{\partial D} y\, dx$$

Can you generalize this to a 3-D domain D using its 2-D boundary ∂D?

This is not a difficult problem: a little pattern-matching and guess-and-check suffices. The difficult parts of this are the ambiguity (there are multiple integrands that will work) and the motivation (why would one want to compute a centroid in this way). For students who think that the Gauss Theorem always goes in one direction – from a difficult double integral to a cleaner triple integral – this is a good problem. For motivation, see the discretized version in Volume 4 Chapter 15 and its uses in medical imaging.

QUESTION 10: Computing induced orientations on boundary curves can be hard to visualize: here is an exercise. Take a sphere of radius 2 centered at the origin, oriented with an outward-pointing normal. From this, remove the three unit-radius solid cylinders about the x, y, and z axes. How many boundary components does the resulting surface have, and what are their orientations?

This is challenging both to explain in words and to draw pictures of: see what students come up with in terms of clever ways to think about this problem.

QUESTION 11: What is Gauss's Theorem good for? Recall Archimedes and the old story of the tub... *Eureka!* Remember the principle that the weight of a floating body equals the weight of the fluid displaced by it. What is the buoyant force on a floating body and how does it relate to weight and volume and *forms*?

This requires a little bit of Physics. A fluid has particles bouncing around at random, applying pressure to a submerged surface that acts orthogonal to the surface itself. The force applied by the fluid is this pressure times the surface area element. Since the floating body does not fly off, this means that the horizontal components of the fluid force all cancel, and the vertical component equals the weight of the body. Assuming the water is of constant density, the magnitude of the pressure force in the z-direction is ρz times the projected area in the (x, y) plane, where ρ is the density. Try to lead students to the observation that the buoyant force field is really a 2-form field $\beta = \rho z\, dx \wedge dy$ (with the proper orientation). Next step: what is the net buoyant force? Ah, that's an integral over the portion of the surface that is submerged. Be sure to emphasize to students that the resulting integral gives the weight of the displaced liquid, and uniform density of the floating body is never assumed. This can be paired with a physical demo & makes a great *application of the divergence theorem. What about the case where the density of the fluid varies? It is not pretty, but it is doable.*

ASSESSMENT PROBLEMS

PROBLEM 1. Integrate directly the 2-form field $\beta = x\, dx \wedge dy + z\, dy \wedge dz$ over the parametrized surface G given by

$$G\binom{s}{t} = \begin{pmatrix} s^2+t \\ s-t^2 \\ s^2-t^2 \end{pmatrix} \quad : \quad -1 \le s \le 1 \;\; , \;\; 0 \le t \le 1$$

PROBLEM 2. Integrate directly the 2-form field $\beta = x\, dy \wedge dz + y\, dz \wedge dx$ over the parametrized surface S given by

$$S\binom{u}{v} = \begin{pmatrix} uv \\ u+v \\ u-v \end{pmatrix} \quad : \quad u^2+v^2 \le 4$$

PROBLEM 3. Use the Gauss [divergence] theorem to compute the flux of the vector field $\vec{F} = (x^3 - z^5)\hat{\imath} + (y^3 + z^4)\hat{\jmath} + (1 + z^3)\hat{k}$ across the full boundary of the solid hemisphere of radius 2 opening up along the z-axis. Assume an outward-pointing normal and that the hemisphere boundary includes the bottom disc in the (x, y) plane.

PROBLEM 4. Use the Gauss [divergence] theorem to compute the flux of the vector field $\vec{F} = (x^3 - z^5)\hat{\imath} + (y^3 + z^4)\hat{\jmath} + (x^5 - y^3)\hat{k}$ across the boundary of the cylinder of radius 2 opening up along the z-axis from $z = 0$ to $z = 4$. Assume an outward-pointing normal.

PROBLEM 5. Consider the 3-d cube with corners at (0,0,0) and (2,2,2). If I tell you that the flux of the vector field

$$\vec{F} = (x^2 - y^2)\hat{\imath} + (x^2 + y^2)\hat{\jmath} + (z^2)\hat{k}$$

across the top face of the cube (where $z = 2$) is equal to 16, then use Gauss' Theorem to compute the flux out of the four sides of the cube where $0 < z < 2$.

PROBLEM 6. Use the Gauss [divergence] theorem to compute the flux of

$$\vec{V} = (xy^2)\hat{\imath} + (yz^2 + x^3)\hat{\jmath} + \left(x^2(z - y^2)\right)\hat{k}$$

across a sphere of radius two centered at the origin. Assume an outward pointing normal.

PROBLEM 7. Let S be the closed surface which forms the boundary of the solid domain $0 \le z \le 9 - x^2 - y^2$. Use the Gauss [divergence] theorem to compute the flux of the vector field $\vec{F} = xy^2\, \hat{\imath} + x^2 y\, \hat{\jmath} + (z - e^{xy})\hat{k}$ across this surface S, using an outward-pointing normal.

PROBLEM 8. Consider the surface S which forms the boundary of the solid domain $x^2 + y^2 - 4 \le z \le 4 - x^2 - y^2$. Compute the flux of the vector field $\vec{F} = -ye^z\, \hat{\imath} + xe^z\, \hat{\jmath} + (x + y + z)\, \hat{k}$ across this surface S, using an outward-pointing normal.

PROBLEM 9. Compute the flux of the vector field

$$\vec{V} = (x - y^2)\hat{\imath} + (z^2 + y^2)\hat{\jmath} + (xy + z)\hat{k}$$

across the cube in 3-D with opposite corner points at $(-1,-2,-3)$ and $(3,2,1)$. Assume an outward pointing normal.

PROBLEM 10. Consider the solid upper hemisphere H given by $x^2+y^2+z^2 \leq 1 \ : \ z \geq 0$. Let S denote the upper hemispherical surface of H. Let D denote the bottom of H : that is, the unit disc in the (x,y) plane. Let $\vec{V}$ denote the vector field $\vec{V} = y^2\hat{\imath} + z^2\hat{\jmath} + (x^2+y^2)\hat{k}$.

A) Compute the flux of $\vec{V}$ across the entire boundary of H; that is, across both S and D, using the usual outward-pointing normals.

B) Compute the flux of $\vec{V}$ across the upper hemisphere S. Use upward-pointing normals.

PROBLEM 11. Use Stokes' Theorem to compute the flux of the curl of the vector field

$$\vec{F} = (x^2z - y)\hat{\imath} + (x - yz^2)\hat{\jmath} + \left(\sqrt{xyz}\right)\hat{k}$$

across the upper hemisphere

$$x^2+y^2+z^2 = 9 \ : \quad z \geq 0$$

oriented via the positive z-axis.

PROBLEM 12. Use Stokes' Theorem to compute the flux of the curl of the vector field

$$\vec{F} = (z^2 - y)\hat{\imath} + (x - z^2)\hat{\jmath} + (x^2+y^2+z^2)\hat{k}$$

across the surface given by the graph of

$$z = x^2+y^2-4 \quad : \quad z \leq 0$$

oriented via the positive z-axis.

PROBLEM 13. Use Stokes' Theorem to compute the flux of the curl of the vector field $\vec{F} = (-zy)\hat{\imath} + (zx)\hat{\jmath} + (xy\cos^2 z)\hat{k}$ across the surface S given by the formula $z = \sqrt{5 - x^2 - y^2}$ and $x^2+y^2 \leq 1$, using an upward-pointing normal (along the z-axis).

PROBLEM 14. Use Stokes' Theorem to compute the flux of the curl of the vector field $\vec{F} = y\hat{\imath} - x\hat{\jmath} + z(x^3 - y^3)\ \hat{k}$ across the surface S parametrized by

$$S\begin{pmatrix} u \\ v \end{pmatrix} = \begin{pmatrix} u \\ v \\ 1 - u^2 - v^2 \end{pmatrix} \quad : \quad u^2+v^2 \leq 1$$

PROBLEM 15. Consider the 1-form field

$$\alpha = (x^2 - 1)dx + (z+1)dy + (2-y)dz\ .$$

Consider the surface S given by the unit sphere in the positive corner of $\mathbb{R}^3$ where $x,y,z \geq 0$. Let γ be the closed curve given by the boundary of S, as shown.

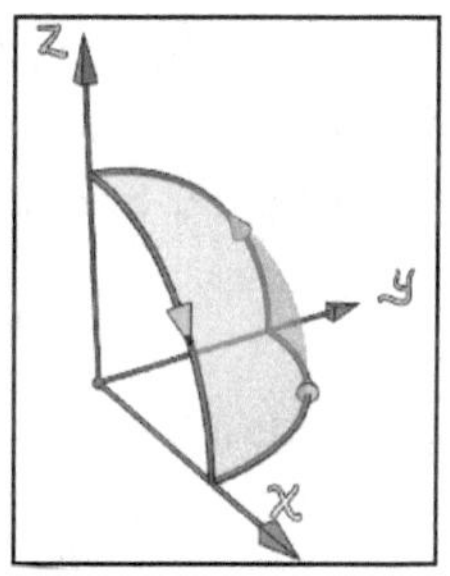

A) Of which vector field is α the work 1-form?

B) Compute the integral of α along γ, the boundary of S.

C) Is α the gradient 1-form of some potential function f? *Why or why not?*

PROBLEM 16. Consider the surface S given by

$$x^2 + y^2 + z^2 = 25 \quad : \quad z \le 0 \quad : \quad x^2 + y^2 \le 4$$

This can be described as a circular region about the *south pole* of a sphere of radius 5 where $z \le -3$. Compute the flux of the curl of the vector field

$$\vec{F} = -yz^2\,\hat{\imath} + xz^2\,\hat{\jmath} + e^{-xyz}\,\hat{k}$$

across this surface S, using an upward-pointing normal (along the $+z$ axis).

PROBLEM 17. Compute the flux of the curl of the vector field

$$\vec{F} = (x^3 - 4y)\hat{\imath} + y^2\hat{\jmath} - z^5\,\hat{k}$$

across the oriented surface given by the following parametrization:

$$S\begin{pmatrix} u \\ v \end{pmatrix} = \begin{pmatrix} u \\ v \\ u^2 - v^2 \end{pmatrix} \quad : \quad u^2 + v^2 \le 4$$

PROBLEM 18. Compute the circulation of the vector field

$$\vec{F} = (4y - xy)\hat{\imath} + y^3\hat{\jmath} + \cos z\,\hat{k}$$

along the loop given by the intersection of the cylinder $x^2 + y^2 = 4$ and the plane $z = x$, oriented any way you prefer.

ANSWERS & HINTS

PROBLEM 1.

$$\int_G \beta = \int_0^1 \int_{-1}^1 (s^2 + t)(-4st - 1) + (s^2 - t)(-2s + 4st)\,ds\,dt = -\frac{5}{3}$$

PROBLEM 2.

$$\int_S \beta = \iint_{u^2+v^2 \le 4} u^2 + v^2\,du\,dv = \int_0^{2\pi}\int_0^2 r^3\,dr\,d\theta = 8\pi$$

PROBLEM 3. by Gauss's Theorem

$$\int_{\partial D} \Phi_{\vec{F}} = \iiint_D 3\rho^2\,dV = \int_0^{2\pi}\int_0^{\pi/2}\int_0^2 3\rho^4 \sin\phi\,d\rho\,d\phi\,d\theta = \frac{192}{5}\pi$$

PROBLEM 4. by Gauss's Theorem

$$\int_{\partial D} \Phi_{\vec{F}} = \iiint_D 3r^2\,dV = \int_0^{2\pi}\int_0^2\int_0^4 3r^3\,dz\,dr\,d\theta = 96\pi$$

PROBLEM 5. by Gauss's Theorem, the flux out of the full cube boundary is

$$\int_{\partial D} \Phi_{\vec{F}} = \iiint_D 2(x + y + z)\,dV = 2\int_0^2\int_0^2\int_0^2 x + y + z\,dx\,dy\,dz = 48$$

as the flux on the bottom face is zero and the top is 16, the side flux is 32

PROBLEM 6. by Gauss's Theorem

$$\int_{\partial D} \Phi_{\vec{V}} = \iiint_D \rho^2\,dV = \int_0^{2\pi}\int_0^{\pi}\int_0^2 \rho^4 \sin\phi\,d\rho\,d\phi\,d\theta = \frac{128}{5}\pi$$

PROBLEM 7. by Gauss's Theorem

$$\int_{\partial D} \Phi_{\vec{F}} = \iiint_D 1 + r^2\,dV = \int_0^{2\pi}\int_0^3\int_0^{9-r^2} r + r^3\,dz\,dr\,d\theta = 162\pi$$

PROBLEM 8. by Gauss's Theorem

$$\int_{\partial D} \Phi_{\vec{F}} = \iiint_D dV = \int_0^{2\pi} \int_0^2 \int_{r^2-4}^{4-r^2} r \, dz \, dr \, d\theta = \frac{64}{3}\pi$$

PROBLEM 9. by Gauss's Theorem, the flux out of the full cube boundary is

$$\int_{\partial D} \Phi_{\vec{V}} = \iiint_D 2 + 2y \, dV = 2V = 128$$

PROBLEM 10. A) by Gauss's Theorem, the flux out of the full boundary is zero, since the divergence of $\vec{V}$ vanishes ; B) integrate r^2 over D to get $\pi/2$

PROBLEM 11. By Stokes' Theorem using the boundary circle where $z = 0$,

$$\int_D \Phi_{\nabla\times\vec{F}} = \int_{\partial D} \alpha_{\vec{F}} = \int_{\partial D} -y \, dx + x \, dy = 18\pi$$

PROBLEM 12. By Stokes' Theorem using the boundary circle where $z = 0$

$$\int_D \Phi_{\nabla\times\vec{F}} = \int_{\partial D} \alpha_{\vec{F}} = \int_{\partial D} -y \, dx + x \, dy + (x^2 + y^2) dz = 8\pi$$

PROBLEM 13. By Stokes' Theorem using the boundary circle where $z = 2$

$$\int_S \Phi_{\nabla\times\vec{F}} = \int_{\partial D} \alpha_{\vec{F}} = \int_{\partial D} 2y \, dx - 2x \, dy + (xy \cos^2 2) dz = -4\pi$$

PROBLEM 14. By Stokes' Theorem using the boundary circle where $z = 0$,

$$\int_S \Phi_{\nabla\times\vec{F}} = \int_{\partial S} \alpha_{\vec{F}} = \int_{\partial S} y \, dx - x \, dy = -\pi$$

PROBLEM 15. A) ; B) By Stokes' Theorem

$$\int_{\partial S} \alpha = \int_S d\alpha = \int_S dz \wedge dy - dy \wedge dz = -\frac{\pi}{2}$$

using the projected oriented area of the surface onto the quarter-disc in the (y, z) plane; this α is not a gradient since its derivative is nonzero (curl of grad is always zero)

PROBLEM 16. By Stokes' Theorem using the boundary circle where $z = -3$,

$$\int_S \Phi_{\nabla\times\vec{F}} = \int_{\partial S} \alpha_{\vec{F}} = \int_{\partial S} -9y \, dx + 9x \, dy + e^{3xy} dz = 72\pi$$

PROBLEM 17. Compute the curl as $4\hat{k}$ directly or use forms to integrate

$$\int_S \Phi_{\nabla\times\vec{F}} = \int_S d\left((x^3 - 4y)dx + y^2 dy - z^5 dz\right) = \int_S 4 \, dx \wedge dy = 16\pi$$

PROBLEM 18. By Stokes' Theorem,

$$\int_{\partial D} (4y - xy)dx + y^3 dy + \cos z \, dz = \int_S (4 - x) \, dx \wedge dy = 16\pi$$

EPILOGUE : BEYOND CALCULUS

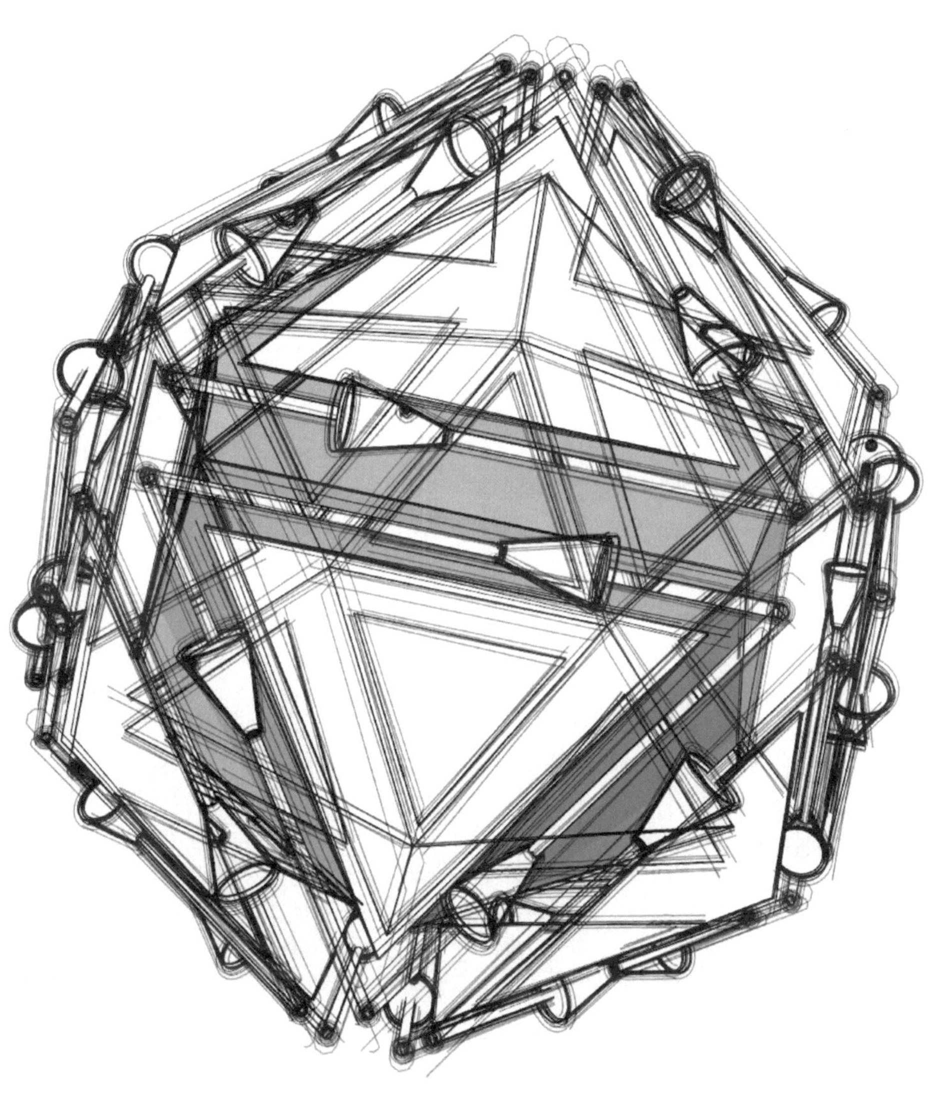

OUTLINE

MATERIALS: Calculus BLUE : Vol 4 : Chapters 13-18

TOPICS:

- ▷ Differential forms in geometric data analysis
- ▷ Differential forms in fluid dynamics
- ▷ Differential forms in electromagnetics
- ▷ Differential forms and calculus on $\mathbb{R}^n$
- ▷ The general Stokes Theorem on $\mathbb{R}^n$
- ▷ Integration by parts via differential forms
- ▷ Differential forms in time series analysis
- ▷ Mathematics beyond Calculus

LEARNING OBJECTIVES:

- ▷ *Inspiration*

PRIMER

There's more to the Story than has been told. For those called to explore past the bounds of this course, there are several chapters of bonus material to give an idea of what differential forms are good for and how the calculus of fields and forms extends beyond 3-D.

[BONUS] FORMS & SPATIAL DATA. One of the best, simplest applications of Green's Theorem is in geometric data analysis in 2-D. Given a domain D in the plane, one can compute its area by adding up changes in the positions of points along the boundary curve $\gamma = \partial D$:

$$A = \int_D dx \wedge dy = \frac{1}{2}\int_\gamma x\,dy - y\,dx.$$

This is of considerable practical use: say if a medical technician traces the outline of a domain on an ultrasound, or a drone traces the boundary about a hazardous interior region in the plane. In practice, the boundary curve is most likely to be discretized – sampled along a cyclic sequence of points. How can the path integral best be approximated?

Consider breaking the boundary curve γ into a sequence of straight-line paths $\{\gamma^i\}$ for $i = 1 \dots N$ with each segment consistently oriented having endpoints from start (x_1^i, y_1^i) to finish (x_2^i, y_2^i). By parametrizing each segment and integrating, one derives the following useful approximation to the area of D:

$$A \approx \sum_{i=1}^{n}\left(\frac{1}{2}\int_{\gamma^i} x\,dy - y\,dx\right) = \frac{1}{2}\sum_{i=1}^{n}\left(x_1^i y_2^i - y_1^i x_2^i\right).$$

This is a subtle formula. What about the case of 3-D data, where a point-cloud approximates the boundary ∂D of a solid body D? Triangulate the surface with a collection of oriented triangles T^i, each with three boundary points located at (x_j^i, y_j^i, z_j^i) for $j = 1 \dots 3$ and $i = 1 \dots N$. In the case of a smooth boundary ∂D, the theorem of Gauss states that the enclosed volume is computed via:

$$V = \int_D dx \wedge dy \wedge dz = \frac{1}{3}\int_{\partial D} x\,dy \wedge dz + y\,dz \wedge dx + z\,dx \wedge dy.$$

This latter surface integral can be discretized into the sums of integrals over the triangles T^i to yield:

$$
\begin{aligned}
V &\approx \sum_{i=1}^{n}\left(\frac{1}{3}\int_{T^i} x\,dy\wedge dz + y\,dz\wedge dx + z\,dx\wedge dy\right) \\
&= \frac{1}{6}\sum_{i=1}^{n}\left(x_1^i y_2^i z_3^i + x_2^i y_3^i z_1^i + x_3^i y_1^i z_2^i - x_1^i y_3^i z_2^i - x_2^i y_1^i z_3^i - x_3^i y_2^i z_1^i\right)
\end{aligned}
$$

This is, as in the 2-D case, a determinant. The above formulae are classics and well-known. Less familiar are analogous formulae for computing centroids and moments of inertia using boundary point data and differential forms. The videotext details an entirely novel application involving estimation of surface area on the surface of the earth (assumed spherical) using boundary point data and differential forms in spherical coordinates: the key is the following computation of surface area on a radius R sphere:

$$
S = \int_D R^2 \sin\phi\, d\theta\wedge d\phi = \int_{\partial D} -R^2\cos\phi\, d\theta\,.
$$

This follows not from Green's or Gauss's Theorem, but from Stokes's, and is just the beginning of a beautiful story of using differential forms in geometric data analysis.

[BONUS] FORMS & FLUIDS. A mathematical *fluid* is a field of particles that flows according to its *velocity field*, $\vec{V}$, the vector field that registers the rate of change of position of each fluid particle as a function of position and time, meaning that $\vec{V} = \vec{V}(x, y, z, t)$. The *vorticity* of the fluid, $\vec{W} = \nabla\times\vec{V}$, is the curl of the velocity and measures how rotational the fluid is at a given location and time. The nicest examples of fluids are the *perfect fluids*, which have no friction (viscosity): these satisfy the *Euler equations* of motion, written using the *material derivative* from the bonus material in Week 6. The velocity field $\vec{V}$ is that of a perfect fluid if its work 1-form and flux 2-form satisfy

$$
\frac{D\alpha_{\vec{V}}}{Dt} = -dh \quad : \quad d\Phi_{\vec{V}} = 0\,,
$$

for some scalar field h, which encodes pressure and other physical aspects of the fluid. The latter equation means that the fluid is *incompressible* (volume-preserving or zero-divergence), and the former equation means that the time-derivative of the work 1-form is a gradient for some potential h.

The fundamental theorems of Weeks 12-14 can be immediately put to work. The first major theorem of perfect fluids – *Kelvin's Theorem* – concerns the circulation of a fluid along a loop. Kelvin's Theorem states that, in a perfect fluid, the circulation C_γ along a loop of particles γ does not change over time. Remember, the particles and thus the loop are changing over time, so that:

$$
\frac{d}{dt}C_\gamma = \frac{d}{dt}\int_{\gamma(t)}\alpha_{\vec{V}} = \int_{\gamma(t)}\frac{D\alpha_{\vec{V}}}{Dt} = \int_{\gamma(t)} -dh = 0
$$

This is a combination of the Euler equation with the Independence of Path Theorem.

The 2nd major result of mathematical fluid dynamics is *Helmholtz's Theorem*, which combines the vorticity 2-form $\omega = d\alpha_{\vec{V}}$ and Stokes' Theorem with the idea of a *vortex tube* that measures how the fluid twists: see the videotext for details.

[BONUS] FORMS & ELECTROMAGNETICS. Vector fields are the traditional language for electromagnetic fields and forces, all connected by the familiar operations of grad, curl, and div, thanks to *Maxwell's equations*. In 3-D, one has the following (time-varying) vector and scalar fields:

- ▷ The *electric field* $\vec{E}$
- ▷ The *magnetic field* $\vec{B}$
- ▷ The *current field* $\vec{J}$
- ▷ The *charge density* ρ

Maxwell equations relates these quantities via the following formulae:

$$\nabla \times \vec{E} = -\frac{\partial \vec{B}}{\partial t} \quad ; \quad \nabla \cdot \vec{E} = \rho \quad ; \quad \nabla \times \vec{B} = \vec{J} + \frac{\partial \vec{E}}{\partial t} \quad ; \quad \nabla \cdot \vec{B} = 0 \,.$$

The theorems of Gauss and Stokes then translate into statements such as:

The flux of the electric field through a closed surface equals the net charge enclosed by the surface.

There is no magnetic charge : the flux of the magnetic field across a closed surface is always zero.

The circulation of the electric field along a loop depends on the flux of how the magnetic field changes over time.

This is fine, but there is a better approach using modern terminology. If one converts the electric and magnetic data into 2-forms and the current/charge into a 3-form, a great simplification arises.

Let $\mathcal{F} = \alpha_{\vec{E}} \wedge dt + \Phi_{\vec{B}}$ be the *Faraday form* and $\mathcal{M} = \alpha_{\vec{B}} \wedge dt - \Phi_{\vec{E}}$ the *Maxwell form*. These are 2-forms on the 4-dimensional space-time with coordinates $(\boldsymbol{x}, t)$. The current-charge form is the 3-form given by $\mathcal{J} = \Phi_{\vec{J}} \wedge dt - \rho\, d\boldsymbol{x}$. Then, Maxwell's equations can be written compactly as

$$d\mathcal{F} = 0 \;;\; d\mathcal{M} = \mathcal{J} \,.$$

The utility of this formulation (besides concision) is the connection to the geometry of spacetime, as $\mathcal{F}$ can be seen as the curvature of the electromagnetic potential. This is the beginning of a much longer story in Physics.

[BONUS] BEYOND 3-D. The above applications to fluids and electromagnetics are both working with time-dependent fields, and the resulting differential forms are on $\mathbb{R}^4$. This prompts the question of how to extend the calculus of form fields from $\mathbb{R}^3$ to $\mathbb{R}^n$. Many features are a simple translation: 0-forms are scalar fields, and 1-forms are generated from the basis $dx_1, dx_2, \ldots, dx_n$. The basis k-forms (for any k) are generated from the wedge product $\wedge$ and a few simple rules involving determinants. A basis k-form $dx_{i_1} \wedge \cdots \wedge dx_{i_k}$ eats an ordered k-tuple of vectors $(\boldsymbol{v}_1, \boldsymbol{v}_2, \cdots, \boldsymbol{v}_k)$ in $\mathbb{R}^n$, stacks them into an n-by-k matrix, then selects k rows (based on the $i_1 \cdots i_k$ indices), and takes the determinant of the resulting k-by-k matrix. Based on this, these basis k-forms are linear functions of each input vector and they are antisymmetric:

- ▷ switching any two dx_* terms reverses the sign; and
- ▷ repeating any two dx_* terms yields zero.

This implies that all k-forms on $\mathbb{R}^n$ for $k > n$ vanish. The basis k-forms are used to build k-form fields. Work is the natural motivation for 1-form fields; flux arises as an $(n-1)$-form field, where, for a vector field $\vec{F} = \sum_i F_i\,\hat{e}_i$ on $\mathbb{R}^n$,

$$\Phi_{\vec{F}} = \sum_{i=1}^{n} F_i\, dx_{i+1} \wedge \cdots \wedge dx_n \wedge dx_1 \wedge \cdots \wedge dx_{i-1} .$$

Form fields can be differentiated, with d taking a k-form field ω to $(k+1)$-form field, $d\omega$, following the usual pattern of $d(f\varepsilon) = df \wedge \varepsilon$ for any basis form ε. The differentiation operator d satisfies a product rule given by:

$$d(\alpha \wedge \beta) = d\alpha \wedge \beta + (-1)^p \alpha \wedge d\beta ,$$

where p is the degree of α; that is, α is a p-form. This asymmetry comes from the fact that for any p-form α and q-form β, $\alpha \wedge \beta = (-1)^{pq} \beta \wedge \alpha$. Derivatives on form fields satisfy the all-important lemma:

$$d^2 = d \circ d = 0 ,$$

generalizing the 3-D results about curl-of-grad and div-of-curl vanishing.

The Fundamental Theorem – *Stokes' Theorem* – uses derivatives and integrals of form fields over generalizations of curves and surfaces in $\mathbb{R}^n$. These generalized surfaces are called *manifolds* and are the subject of a great deal of more advanced Mathematics. For the present, the term "*k-dimensional domain*" will refer to such an object, a generalized surface that locally "looks like" $\mathbb{R}^k$ in the same way that a curve is everywhere locally a line, or a surface is locally a plane – no kinks or singularities.

Integration of form fields proceeds following the pattern in $\mathbb{R}^3$: 1-form fields are integrated over 1-dimensional domains; 2-form fields are integrated over 2-form fields; k-form fields are integrated over k-dimensional domains. Such domains are parametrized by functions $G: \mathbb{R}^k \to \mathbb{R}^n$ of k parameters $t_!, \ldots, t_k$. To integrate a k-form field ω over G one feeds the columns of the derivative $[DG]$ into ω to obtain a scalar field, which is integrated over the domain in $\mathbb{R}^k$. The usual Change of Variables application reveals that the integral is independent of the parametrization, so long is the orientation is consistent. With these tools in place, one can state the Fundamental Theorem:

THE GENERALIZED STOKES' THEOREM

$$\int_{\partial D} \omega = \int_D d\omega .$$

This holds for any k-form field ω defined on an oriented $(k+1)$-dimensional domain D with oriented k-dimensional boundary ∂D.

This is the end of the Story, but not the last word. This Fundamental Theorem is as useful as it is beautiful, and it is very beautiful. The applications are not obvious and can require additional background; to that end, we close with a few elementary applications of forms in higher dimensions.

[BONUS] PARTS. The only integration technique we dwelt on in this course was substitution – this was the content of the Change of Variables Theorem in Week 11. Among the other integration techniques that you have seen in the past, there is one that is directly related to Stokes' Theorem: integration by parts. As with the single variable version, one integrates the product formula for derivatives: for α a p-form field and β a q-form field; for D a $(p+q+1)$-dimensional domain with $(p+q)$-dimensional boundary,

$$\int_{\partial D} \alpha \wedge \beta = \int_D d(\alpha \wedge \beta) = \int_D d\alpha \wedge \beta + (-1)^p \alpha \wedge d\beta \,.$$

This has numerous applications in partial differential equations and analysis in the guise of so-called *Green's Identities*.

[BONUS] GEOMETRIC OPTICS. Why would one care about form fields in higher dimensions? One can argue for 4-D calculus based on space-time, as in the case of fluids and electromagnetics; higher dimensions are not as obvious. One class of examples that is both physical and easily imagined comes from optics and the study of light rays, used in both Cosmology and Computer Graphics. Consider a ray of light – a straight line in 3-D space, coordinatized as a z-axis, with an orthogonal (x, y) plane that coordinatizes translations of the light ray. Changing the directions of light rays is controlled by two angles (φ, ψ) which rotate in the (x, z) and (y, z) planes respectively. Then, the space of rays is a 5-dimensional space.

Differential forms provide a convenient language for working with the geometry of light rays. One can describe a distribution of light rays using a *Hamiltonian* scalar field $H(x, y, \varphi, \psi)$ via the following 2-form field:

$$\Omega = d\varphi \wedge dx + d\psi \wedge dy + dz \wedge dH \,.$$

The *brightness* of a ray distribution is a 4-form field $B\Omega \wedge \Omega$ for B a scalar field. The integral of this brightness 4-form over a 4-D domain that represents a lens gives the *throughput* or *étendue* of the lens. Other fundamental concepts in geometric optics overlap with Hamiltonian mechanics and dynamics, for which differential forms are foundational.

[BONUS] STOKES' THEOREM & TIME SERIES DATA. The following is a novel application of differential forms and Stokes' Theorem in $\mathbb{R}^n$ to data analysis, based on work with Yuliy Baryshnikov. Consider a collection of time-series data – one can think of N signals $x_i(t)$ for $i = 1 \ldots N$ as functions of time t.

Assume that the signals x_i are measuring phenomena that are time-periodic, such as the swinging of a pendulum or the boom-bust business cycle or any number of biological signals based on circadian rhythms. If we have several such signals with the same period, $x_i(t+P) = x_i(t)$, then one might want to know whether one is a *leading* or a *lagging* indicator – this is especially useful in Economics and in Biology.

For purely periodic functions, there are many ways to discern this order (*harmonic analysis* is one such). Those methods are, however, very sensitive to time axis reparameterizations. Many real-life phenomena are *cyclic* without being rigidly *periodic*. Cardiac rhythms, musculo-skeletal movements exercised during a gait, population dynamics in closed ecosystems, business cycles, neural responses, and more are examples of cyclic yet aperiodic processes.

The key insight is that each pair of (roughly) time-periodic signals $\left(x_i(t), x_j(t)\right)$ traces out a closed curve in the $\left(x_i, x_j\right)$ plane whose oriented area reveals leading or lagging behaviors and the degree thereof. This oriented area is measure by the 2-form $dx_i \wedge dx_j$ in the full signal space $\mathbb{R}^n$. A simple application of Stokes' Theorem means that one can compute this oriented projected area by integrating over the boundary curve:

$$A_{ij} = \int_D dx_i \wedge dx_j = \frac{1}{2}\int_{\partial D} x_i\, dx_j - x_j\, dx_i.$$

The curve is automatically time-parametrized by $x_i(t)$ and $x_j(t)$.

For temporally discretized data (such as might occur in experimental or social sciences) an estimate based on piecewise-linear paths as at the beginning of this Epilogue is effective and robust with respect to nonuniformities in sampling of points. One can assemble all these path integrals into an antisymmetric lead-lag matrix $A = [A_{ij}]$ with $A_{ji} = -A_{ij}$ that correlates signed lead-lag behaviors, all independent of time parametrization, thanks to Stokes's Theorem.

[BONUS] BEYOND CALCULUS. Differential forms point the way to a universe of mathematical subjects of current and enduring interest in research:

▷ In *differential geometry*, forms are the language for describing all types of curvature on geometric (Riemannian) manifolds.

▷ In *algebraic topology*, forms give a fundamental example of *cohomology* and Stokes' Theorem gives a precise mechanism for *Poincaré duality*.

▷ In *real analysis* and *differential equations*, forms and Stokes' Theorem provide an elegant generalization of *Green's Identities*.

▷ In *complex analysis*, forms and Green's Theorem are the basis for deep integration results, including *Cauchy's Integral Theorem*.

▷ In *algebraic geometry*, forms are a crucial ingredient in *Hodge Theory* and many other approaches to classifying regular and singular behavior in solutions to polynomial equations.

There is so much more that you can learn and do with the background you now possess.

CALCULUS BLUE GUIDE

1st edition, corrected

Published by Agenbyte Press
Jenkintown PA, USA

ISBN 978-1-944655-07-5

ABOUT THE AUTHOR

Robert Ghrist (Ph.D., Cornell, Applied Mathematics, 1995) is the Andrea Mitchell PIK Professor of Mathematics and Electrical & Systems Engineering at the University of Pennsylvania. He is a recognized leader in the field of Applied Algebraic Topology, working in sensor networks, robotics, signal processing, data analysis, optimization, and more. He is an award-winning researcher, teacher, and expositor of Mathematics and its applications.

He is the author of *Elementary Applied Topology* and creator of several mathematics video-text series on YouTube, including
Calculus BLUE
Calculus GREEN
Applied Dynamical Systems

Ghrist has been an invited speaker at two International Congresses of Mathematicians: once (Madrid 2006) for research and once (Seoul, 2014) for education. Ghrist is a dedicated expositor and communicator of Mathematics, with teaching awards that include the MAA James Crawford Prize, Penn's Lindback Award, and the S. Reid Warren award in Engineering at Penn.

In his spare time
he publishes mathematical art and animation
under the moniker *colimit*

colimit.eth.xyz
objkt.com/profile/colimit/created

Made in United States
Troutdale, OR
09/14/2024

22798095R00111